FOCKE-WULF TA 152

Dietmar Hermann

Focke-Wulf Ta152

The Story of the Luftwaffe's Late Variant High-Altitude Fighter

Schiffer Military History
Atglen, PA

For Regina and Marc Benedict

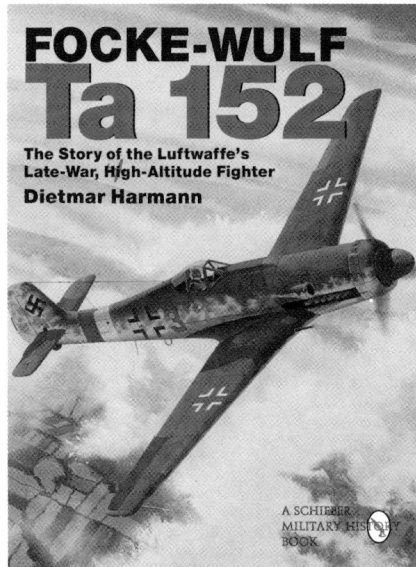

Book Design by Ian Robertson.

Copyright © 1999 by Schiffer Publishing, Ltd.
Library of Congress Catalog Number: 99-60563.

Translation from the German by David Johnston.

This book was originally published under the title *Focke-Wulf Ta152: Der Weg zum Höhenjäger* by Aviatic Verlag.

Printed in China.
ISBN: 0-7643-0860-2

We are interested in hearing from authors with book ideas on military topics.

Published by Schiffer Publishing Ltd.
4880 Lower Valley Road
Atglen, PA 19310 USA
Phone: (610) 593-1777
FAX: (610) 593-2002
E-mail: Schifferbk@aol.com.
Visit our web site at: www.schifferbooks.com
Please write for a free catalog.
This book may be purchased from the publisher.
Please include $3.95 postage.
Try your bookstore first.

In Europe, Schiffer books are distributed by:
Bushwood Books
6 Marksbury Road
Kew Gardens
Surrey TW9 4JF
England
Phone: 44 (0)181 392-8585
FAX: 44 (0)181 392-9876
E-mail: Bushwd@aol.com.

Try your bookstore first.

Contents

Foreword

Development of the TA 152 in all its variants was based on the fundamental design of the Focke-Wulf Fw 190 A series powered by the BMW 801 radial engine. The design of the Ta 152 by Prof. Kurt Tank developed very quickly and so differed from the original design that it became a completely new fighter aircraft, the Ta 152.

As with other innovative developments by the German aviation industry in the final years of the war, the comment too little and too late also applied to the Ta 152. Production of the Ta 152 H, Germany's first high-altitude fighter, was initiated in the late autumn of 1944 but was halted soon afterwards by the total collapse of the armaments industry. As a result, only a very few Ta 152s reached the operational units of the *Luftwaffe*. The pilots who had the chance to fly the Ta 152 in training and combat missions were enthusiastic about the machine's excellent handling qualities. The Ta 152 was at least equal to the opposing Allied fighter aircraft and for the first time in many months German pilots were again in a position to survive aerial combat with Mustangs and Spitfires. In April 1945 the last Ta 152s were concentrated in the expanded *Stabsschwarm* of JG 301 and fought on to the bitter end. Obfw. Willi Reschke, one of the last Ta 152 pilots, remembers that time and said of it: "These were probably the *Stabsschwarm*'s most difficult days and we were only able to survive them because the Ta 152's climbing and turning abilities were so exceptionally good. I had experienced all the highs and lows of both *Jagdgeschwader* 301 and 302 since June 1944, and I would have been happy if I had always had available such a machine during my missions and air battles."

But before the first Ta 152 reached the *Luftwaffe* units, prototype aircraft had to be built and tested to ensure that the new fighter aircraft was equal to the severe demands of the air war. In charge of this testing was the director of Focke-Wulf's Prototype Testing Department, chief test pilot Dipl.Ing. Hans Sander. As an eye witness and last surviving Focke-Wulf test pilot, I am especially grateful to him for his kind help and support for this book. As well, I would like to express my appreciation to the many aviation enthusiasts and to the aviation archive of the Daimler-Benz Aerospace AG in Bremen for their help and support in the creation of my book on the Ta 152.

Dipl.Ing. Dietmar Hermann
Dortmund, June 1998

Introduction by Chief Test Pilot Hans Sander on Testing of the Ta 152

In the area of airframes the German aviation industry, meaning Focke-Wulf, had done much to produce a good high-altitude fighter (extended wingspan). What it had neglected to do, in spite of my exhortations, was develop a reliable pressurized cockpit, which, thank God, played no role in operations.

The German aero-engine industry failed to produce reliable high-altitude engines, and in particular turbo-superchargers, in time, which was due in large part to material shortages. Prototype construction, prototype testing and prototype trials really did everything in this difficult time. Toward the end of the war we test pilots had to provide proof of the aircraft's structural integrity in the air, because stress tests on the ground were no longer possible on account of the general situation.

The question of whether it would not have been less risky to build a high-altitude fighter based on the Fw 190 D-9 with GM-1 injection can only be answered with no. The Fw 190 D-9, which went into production a few months before the Ta 152 H, could never have played the role of that aircraft. Furthermore one could not compare MW 50 injection (usable only at altitudes where excess boost pressure was available) with GM-1, which injected oxygen at any altitude. We, the prototype test pilots, really did everything to test the Ta 152 as quickly as possible in spite of incursions by Allied fighters and bombers.

I am amazed at all the information the author has succeeded in amassing, the amount of work that he has done. I wish this book much success.

Hans Sander
Chief engineer and chief pilot of
Focke-Wulf's main prototype testing department

Development of the Focke-Wulf Fw 190 into the Ta 152 With In-line Engine

When the first prototype of the Fw 190, the V 1, serial number 0001, registration D-OPZE, made its first test flight on 1 June 1939 with test pilot Hans Sander at the controls, no one could yet imagine what a successful and extraordinary fighter aircraft this machine would become.

Once the most significant shortcomings had been eliminated, in August 1941 the first production examples of the Fw 190 A-1 were delivered to *Jagdgeschwader* 26 based on the Channel Coast. Although the Fw 190 A's 14-cylinder BMW 801 radial engine gave it a beefy appearance, the aircraft proved to be a dangerous opponent for the RAF's Hurricanes and Spitfires for a long time.

But it soon became clear to Focke-Wulf's chief designer Kurt Tank that development of the Fw 190 powered by the BMW 801 radial engine would have to continue in order to increase its performance. By the end of 1942 it had become apparent that there would be no significant improvement in the BMW 801 radial engine as installed in the Fw 190 A series in the near future.

Kurt Tank therefore called for the construction of variants of the Fw 190 powered by more powerful engines, the Junkers Jumo 213 and the Daimler Benz DB 603.

Alternative designs were produced in order to achieve a result as quickly as possible. The later developments were based on these so-called Ra (mathematical prospectus) designs. The designs Fw 190 Ra 1 to Ra 6 were presented to the RLM's Technical Office in the spring and were based on the new Junkers Jumo 213 and Daimler Benz DB 603 liquid-cooled, in-line engines.

Junkers developed the Jumo 213 engine from the Jumo 211 power plant, retain-

ing a displacement of 34.97 liters. As a result of increased revolutions the Jumo 213 A produced 1,750 H.P. (1287 kW) for take-off. A similar output was achieved by the Daimler Benz DB 603 engine, a developed version of the DB 601 motor with an increased displacement of 44.5 liters. Both engines were water-cooled, twelve-cylinder, in-line engines, each with two banks of six cylinders in an inverted-vee configuration.

Not long afterwards, in May 1943, Focke-Wulf presented the design of the Ta 152 to the Technical Office (TA) of the Ministry of Aviation (RLM) for the first time. The new project was so far removed from the original design of the Fw 190 as a result of changes and simplifications that in a conference held by the State Secretary on 17 August 1943 it was generally assigned the new designation Ta 152.

Focke-Wulf Fw 190 V 1, registration D-OPZE, *Werknummer* 0 001, with engine running prior to first flight on 1 June 1939.

9

Testing the Fw 190 V 1, which arrived at Rechlin for the first time on 9 July 1939; from left to right: Generalingenieur Lucht, *Generaloberst* Udet, Dipl.Ing. Francke.

Moreover, this was the first time that designer Kurt Tank was permitted to use his name (Ta) in the designation. It was a special honor. Further designs bearing the Ta designation followed, for example the all-wood Ta 154 twin-engined night fighter and the Ta 183 single-engined jet fighter.

Kurt Tank took new paths in the development of the Ta 152. In order to avoid major redesign and the associated loss of time, the Ta 152 received a circular nose radiator for the engine oil and coolant. At the same time, this design choice led to stability problems in the cooling system in the initial

phase. Design work went ahead at Focke-Wulf under the designation Ta 152 Ra 1 (Ta 152 V 1) for the standard fighter and Ta 152 R2 for the high-altitude fighter. In order to accelerate design work on the Ta 152, which was conceived as a fighter and fighter-bomber, on 8 October 1943 an application was made to the RLM for development priority for the Ta 152 A standard version, as it was now designated. This application was initially rejected. The same thing happened to the Ta 152 A as to the Messerschmitt Me 262 A. The RLM was avoiding the risk of deciding in favor of a new type,

The first Fw 190 A-1s, still powered by BMW 801 C engines, on the apron at Bremen, 1941.

so as to avoid endangering the ongoing production of the Fw 190 A and Bf 109 G. Focke-Wulf subsequently resubmitted the application for development priority on 20 December 1943. At a conference between Focke-Wulf and representatives of the RLM on 13-14 January 1944, this second application, for the Ta 152 A, Ta 152 H (high-altitude fighter) and Fw 190 D (Fw 190 with Jumo 213 A engine) was at first not approved. One reason for this fatal hesitation on the part of the RLM was its reluctance to decide for or against construction of the Me 209 A, which was supposed to replace the obsolescent Bf 109. The first flight by the Me 209 (Me 209 V 5, registration SP + LJ) took place on 3 November 1943 with test pilot Fritz Wendel at the controls, almost at the same time as the maiden

flight of the Focke-Wulf Fw 190 V 20, the second prototype of the Ta 152. Not until the Me 209 was removed from the *Luftwaffe's* procurement program was the way opened for the Ta 152, but then only for the Ta 152 high-altitude fighter. The Ta 152 A was struck off, nevertheless the type remained part of the *Luftwaffe*'s planned fighter equipment until the summer of 1944.[1] Consequently production of the Ta 152 A was limited to three prototypes. Service use of the Ta 152 H high-altitude fighter proved that the concept of the Ta 152 was valid. For the first time in a long while the *Luftwaffe* brought to the front a machine that was the technical equal of the Allied fighters. It was prevented from becoming a success only by the collapse of production and the end of the war.

First pre-production Fw 190 A-0s with BMW 801 C engines at the final assembly works in Bremen.

[1] Technical Office memo GL/C-B GL/C No. 1563/44 dated 3 June 1944.

11

The Ta 152's Test Pilots

Professor Tank, Focke-Wulf designer, in conversation with Hptm. Walter Nowotny, who in 1944 formed the first jet fighter unit equipped with the new Me 262 A-1a.

The success of an aircraft was to a large extent dependent on the experience and qualities of the test pilots whose job it was to check out aircraft destined for the front. So it was at Focke-Wulf. Focke-Wulf's chief designer Kurt Tank insisted that the pilots have a good technical background as well as be good flyers. Kurt Tank himself insisted on test flying the aircraft produced under his direction. For example, on 14 April and 29 May 1944 he flew the Fw 190 V 21 (*Werknummer* 0043, registration TI + IH) and on 13 December 1944 the Ta 152 C

V 6 (WNr. 110 006, registration VH + EY).

On several occasions he risked his life by doing so, as at the end of 1944. Tank was on his way to an important conference in Cottbus. Soon after taking off from Langenhagen he was advised, "*Two Indians over the Garden Fence,*" which meant that two enemy fighters were heading straight for Langenhagen airfield. Then two Mustangs appeared directly behind him, hoping for easy prey. Tank had no other choice and applied full power. He engaged the MW 50 injection and quickly pulled away from the Mustangs, which had been closing rapidly, until they were no more than two dots on the horizon. As always, the aircraft Kurt Tank was flying was unarmed, since even at that stage of the war he wished to remain a civilian. He nevertheless made a lasting impression on the two dumbfounded Mustang pilots, who must have wondered which German fighter had such reserves of power.

The best-known Focke-Wulf test pilot was Hans Sander. Until the end of the war he was chief engineer and chief pilot of the "Prototype Testing Department." Hans Sander was involved in the Fw 190 program from the very beginning. On 1 June 1939 he made the first flight in the Fw 190 V 1 powered by the new BMW 139 radial engine and later also flew the latest versions of the Fw 190 powered by the Jumo 213 and DB 603 in-line engines. He was also at the controls for the first flight of the turbosupercharged Fw 190 V 18 "*Höhenjäger 2*" on 20 December 1942. The Ta 152 represented the apex of piston-engined fighter development by Focke-Wulf.

Hans Sander also played a significant role in flight testing the Ta 152 and carried out acceptance tests on the first Ta 152 H fighters produced at Cottbus. For example, on 21 November 1944 he flew the first Ta 152 H-0 (WNr. 150 001, CW + CA), on 29 November WNr. 150 002 (CW + CB), and on 3 December 150 003 (CW + CC). Hans Sander remembered: "I had to put the first production machine down on its belly away from Cottbus because while climbing out after takeoff the engine suddenly stopped receiving fuel. A hydraulic valve had so*me-how* been installed in the fuel line. I received a bottle of schnapps, hard to come by in those days, as compensation. Everything was o.k. with the second machine." Every flight was a calculated risk for a test pilot, including Hans Sander. On numerous occasions he experienced and survived critical situations such as belly landings, fires in the cockpit or flutter at high speeds. Hans Sander remained associated with aviation after the war. He made his last flight in a glider in 1980 at the advanced age of 72.

Although previously a complete unknown, test pilot Bernhard Märschel carried out many test flights in the latest Ta 152s and Fw 190s powered by the Jumo 213 and DB 603 in-line engines. He was at the controls for the first flight of the new Ta 152 C V 6 (VH + EY, WNr. 110 006) at Adelheide on 12 December 1944. This flight also marked the resumption of development work on the Fw 190/Ta 152 series equipped with Daimler Benz engines after a hiatus of more than two years. Kurt Tank had made repeated calls for such work two years earlier but had been turned down by senior officials in the RLM. Bernhard Märschel also remained active in aviation after the war as a glider pilot.

Test pilot Friedrich Schnier achieved the highest altitude reached during the Ta 152 H test program. Flying the Fw 190 V 29/U1, on 20 January 1945 Schnier reached an altitude of 13,654 meters, demonstrating the Ta 152's suitability for the high-altitude fighter role. In addition to the Ta 152, Friedrich Schnier also test-flew the latest variants of the long-nosed Fw 190 D-11/12/13 and the twin-engined Ta 154. Another

experienced test pilot was Werner Bartsch, who was at the controls for the first flights of the Ta 152 V 4 and V 5. A serious accident in the Ta 154 V 9 on 18 April 1944 brought his career as a test pilot to an abrupt end.

Another Focke-Wulf test pilot known to have participated in the Ta 152 test program was Flug*kapitän* Alfred Thomas, who lost his life in the crash of a Ta 152 H (Fw 190 V 30/U1) on 23 August 1944. The aircraft crashed while on approach to Adelheide following an engine fire at altitude. All that is known for certain is that Thomas

Hans Sander, Focke-Wulf's chief test pilot, climbing out of the cockpit of a Focke-Wulf Fw 190 A (WNr. 0 410). Sander played a major role in the success of the Fw 190 and later the Ta 152.

Focke-Wulf test pilots; from left to right: Hans Kampfmeier, Werner Finke, Rolf Mondry, Hptm. Nowotny, Alfred Motsch.

Well-earned break on the airfield boundary; from the right: test pilot Werner Bartsch, next to him Bernhard Märschel, who played a significant role in the testing of the inline-engined Fw 190 D and Ta 152, 2nd from left Wallenhorst, chief controller of prototype test flying.

could have bailed out but did not do so in an attempt to save the Ta 152 H. He almost made it, but then went down with the aircraft from low altitude.

Other Focke-Wulf test pilots like Rolf Mondry and Hans Kampmeier were out of action by the time the Ta 152 was born. Rolf Mondry lost his left arm in a strafing attack

Test pilot Bernhard Märschel (facing away) in conversation with Georg Kohne, a member of the proto-type test unit, who is-sues final instructions before a test flight. Kohne was seriously injured in a bombing raid against KG 40 in Bordeaux.

on Langenhagen and Dipl.Ing. Hans Kamp-meier, whom Kurt Tank described as a Jack of all trades, landed at Hannover-Vahren-wald after a mission with a shot-up canopy and survived.

This book is recognition of their ef-forts. Without them the Ta 152, especially the so badly-needed Ta 152 H high-altitude fighter, would never have been ready for production.

Bernhard Märschel enjoys a cigarette break between test flights.

Adelheide—Focke-Wulf Branch Plant 8—Prototype Construction

The military airfield at Adelheide was constructed in 1936. The first unit to occupy this airfield was III *Gruppe* of *Kampfgeschwader* 27 "*Boelcke*". When the war began this *Gruppe* moved to Brandenburg and took part in the first attacks on Warsaw on 1 September 1939. III *Gruppe* returned briefly to Adelheide after the Polish Campaign before being transferred to the Channel Coast. Adelheide reached its military high point as a staging base for the invasion of Denmark, when a steady stream of aircraft took off and landed there during the night of 8-9 April 1940. This was repeated one month later on 9-10 May when the campaign against France, Belgium and Holland began. That night elements of KG 4 took off from there on raids intended to soften up Amsterdam, Rotterdam and The Hague in preparation for airborne landings.

Following the Battle of Britain activity at Adelheide dwindled further and after the spring of 1941 practically no more military sorties were flown from there. After the departure of the flying units the hangars and some of the barrack blocks in Adelheide stood empty.

Then in June 1941 Focke-Wulf began setting up its Branch Plant 8 in Adelheide. The new factory was solely responsible for the construction of prototypes and began with a complement of 1,500 employees. At the end of 1941-beginning of 1942 this figure rose to 1,900 employees, after which it inexplicably dropped to less then 1,200 workers by the end of the war.

Almost all of the prototypes of the Fw 190 and of course also those of the new Ta 152 were built in the Adelheide factory. Then, once they had successfully completed their first flights, they departed for Langenhagen, where Focke-Wulf conducted technical testing. But Adelheide was repeatedly used again by Lu*ftwaffe* units. One such unit was I/JG 6, which on the morning of 1 January 1945 took part in "Operation *Bodenplatte*" together with the rest of the *Geschwader*. The military end began with

Two Fw 200s on the airfield in Adelheide, which from June 1941 was used by Focke-Wulf for the construction of prototypes.

the disbandment of III/JG 26 "*Schlageter*" on 25 March 1945.

On 14 April 1945 the airfield was prepared for demolition and was partially destroyed beginning on 16 April. On the morning of 19 April 1945 the Scottish 51st Highland Division took the area of the airfield after fierce resistance.

Fw 200 four-engined transport on the airfield in Adelheide; the airfield buildings are just visible in the background.

Development of the Heavy Fighter
The Focke-Wulf TA 152 A and B Series

The development and design of the TA 152 was essentially based on the following requirements issued by the RLM:

1. Installation of the Jumo 213 A standard power plant in the Fw 190 A production airframe with the minimum possible modifications and the maximum possible use of existing jigs and tools, while ensuring the possibility of installing the Jumo 213 E and DB 603 G.

2. Bolstering the centrally-mounted armament, in particular installation of an engine-mounted MK 103 or MK 108. The requirement for minimum possible modifications also applied here.

3. Installation of larger 740 x 210 wheels to cope with the increased takeoff weight resulting from the above requirements.

The overall design of the TA 152 also allowed for the possible use of all three engines under consideration for this type. The version powered by the Jumo 213 A engine was designated the Ta 152 A, while the variant with the Jumo 213 E was the Ta 152 B. The Daimler Benz DB 603 G engine was considered an alternate solution for both versions. But since the Jumo 213 A was not designed for the specified centrally-mounted weapon, the Junkers Jumo 213 C, which had been developed into a standard power plant, would have had to have been installed instead.

Significant features of the overall design were the change from the previously fully electric to a hydraulically activated undercarriage,[1] the larger fin and rudder taken from the "*Höhenjäger* 2" development, and the introduction of a 0.5-meter-long insert in the rear fuselage. The purpose of the latter was to compensate for the altered center of gravity resulting from the longer engine and added weapons in the forward fuselage, and ensure stability about the yaw axis at higher altitudes. Another significant change was the flame damper installation planned from the outset for the Ta 152 A.[2] This was something completely new for a fighter aircraft and still had to be tested. The flame damper installation prevented the pilot from being blinded by the exhaust flames, a problem encountered in single-engined night fighter operations.[3] The use of flame dampers was supposed to make the Ta 152 A suitable for both the day and night fighter roles. But then on 18 April 1944 the requirement for the flame damper installation was dropped (stipulation by the General Staff). An 85-liter GM 1 tank[4] was planned for the Ta 152 to provide increased performance at high altitude. At an average consumption of 100 g/sec this allowed GM 1 to be used for approximately 17 minutes. The supercharger air intake was improved aerodynamically and was extended to the rotational axis of the radiator gills, which then became an identifying feature of all three prototypes.

The Ta 152 A-1 was supposed to be equipped with the MK 108 cannon and the Ta 152 A-2 with the MK 103. The same applied to the Ta 152 B-1 and B-2.

[1] The Fw 190 V 1 and V 2 used hydraulically-activated undercarriages in 1939.

[2] The flame damper installation developed by Focke-Wulf was later installed on the few Ta 154 night fighters powered by the Jumo 213 A.

[3] The "Wild Boar" night fighter method.

[4] GM 1: Injection of an oxidizer (nitrous-oxide) for increased engine performance at high altitudes.

Technical Specification No. 270 for the Ta 152 A/B Dated 16 December 1943

General Information:

Purpose:	Single-seat fighter, fighter-bomber
Construction:	Single-engined, low-wing cantilever monoplane with hydraulically-retractable undercarriage
Strength:	Maneuvering load coefficient nA = 6.5 at a design takeoff weight of 4 400 kg
Power Plant:	Jumo 213 A, Jumo 213 E or DB 603 G

Dimensions:

Wing area	19.6 m2
Wingspan	11 m
Aspect ratio	6.17
Vertical fin area	1.77 m2
Horizontal stabilizer area	2.89 m2
Maximum length	10.784 m
Maximum height	3.360 m

Normal takeoff weight:
4 460 kg with Jumo 213 A (TA 152 A)
4 620 kg with Jumo 213 E (TA 152 B)
4 520 kg with DB 603 G (alternate solution)

Construction Materials: Dural and steel

Armament:
Normal: 5 weapons
2 MG 151 in fuselage with 150 rounds per gun
2 MG 151 in wing roots with 175 rounds per gun
1 engine-mounted MK 108 with 85 to 90 rounds
or 1 engine-mounted MK 103 with 75 to 80 rounds
Additional: 2 weapons
2 MK 108 in wing roots with 55 rounds per gun
or 2 MG 151 in outer wings with 140 rounds per gun
or 2 MK 103 under the wings with 40 rounds per gun

Fuselage:

Essentially the following changes were made to the fuselage:

A lengthening of the forward fuselage by 772 mm was necessary as a result of the greater space required by the engine-mounted MK 103 and the two MG 151s in the fuselage. In order to minimize the procurement of new jigs and tools to an absolute minimum, the fuselage extension was bolted directly to the existing engine attachment points. The wing, which was moved forward 420 mm for center of gravity reasons, was attached in the center of the insert. At the same time, repositioning the wing made it necessary to relocate the rear spar junction and the corresponding fuselage bulkhead. The resulting revised location of the forward fuel tank made it necessary to redesign the fuel tank compartment cover and fuselage sides in the affected area. In order to avoid having to accept a reduction in stability, especially directional stability as a result of the lengthened engine compartment, the aft fuselage was lengthened through the insertion of a cylindrical section 0.5 meters in length. The latter also served to accommodate the oxygen bottles and compressed air bottles for the engine-mounted cannon which were moved aft for cg reasons. The increased fuselage moment resulting from the lengthened fuselage made it necessary to strengthen the frame assembly. This strengthening was accomplished by fitting steel extrusions instead of the Dural extrusions previously used.

Undercarriage

The undercarriage leg including shock strut and mounting was retained. The former electric drive was changed to a hydraulic system. Larger 740 x 210 wheels were installed on account of the aircraft's increased takeoff weight.

Wing

The wing was retained unmodified; only the landing flaps were changed to hydraulic operation. The vertical tail used was the larger unit (1.77 m²) which was supposed to be used in combination with the standard rear fuselage including standard tailwheel on all Focke-Wulf fighter types at the time production of the Ta 152 began.

Control System

The control system remained essentially unchanged, however the more forward engine position and the fuselage extensions resulted in some changes to the linkages.

Wing Assembly

In order to provide propeller clearance from the larger wheels, it was necessary to move the latter outboard by 250 mm. This was made possible by increasing the span of the existing wing from 10.5 to 11 meters, which was achieved by inserting a 0.5-meter spar section in the center of the wing. In turn this made it necessary to redesign the wing-fuselage junction. As a result of the greater wingspan and the increase in the product of n x G, it became necessary to strengthen the skinning in the area of the inner wing.

Power Plant

Installation of the following engines in the TA 152 was anticipated:

1   Jumo 213 A (C) standard fighter engine with Junkers VS 9 variable-pitch propeller, which was developed by Focke-Wulf and which was first used on the production Ta 152. Production began with the flame damper installation

developed by Focke-Wulf. If flight testing revealed a considerable loss in airspeed caused by the flame dampers compared to fighters with standard exhausts, an immediate change to a normal exhaust configuration was to follow for the day fighter version.

2   Jumo 213 E standard fighter engine with Junkers VS 9 or VS 19 variable-pitch propeller (the latter was under development by Junkers at that time). On account of its considerably improved high-altitude performance compared to the Jumo 213, it was planned to switch to the Jumo 213 E for the Ta 152 as soon as the engine was available. In contrast to the Jumo 213 A standard engine, no flame damper installation was initially planned.

3   DB 603 G standard fighter engine, which was foreseen as a alternate solution to the Jumo 213 A. At the time the DB 603 G standard engine was being developed by the Daimler Benz company in cooperation with Focke-Wulf. The power plant was equipped with normal exhausts.

Fuel System

It was possible to retain the 233-liter forward fuel tank from the Fw 190 A series unchanged, however it was moved forward in keeping with the revised wing position. As a result it was possible to increase the capacity of the rear tank by 70 liters to a total of 362 liters. This increased standard tankage to a total of 595 liters. The tanks were the protected type with the following wall thicknesses: sides and bottom 16 mm, top 12 mm. For increased range the following auxiliary tank installations were planned:

1   Metal-shielded 115-liter tank in rear fuselage with a tank wall thickness of 14 + 2 mm. The tank had the same external dimensions as the GM 1 tank described below and was interchangeable with it.

2   Unprotected 300-liter tank under the fuselage. The tank was mounted on the external stores rack built into the wing center section. Installation of an 85-liter GM 1 tank in the fuselage was planned for increased performance at high altitude. Mounting was similar to that of the 115-l auxiliary fuel tank.

Endurance with GM 1 was approximately 17 minutes at an average consumption of 100 g/sec (for increased performance figures see spec sheet p. 29)

Lubrication System

The oil tank, with a total capacity of 64 liters, was installed on the right side of the forward fuselage extension next to the engine cannon. It was a simple sheet steel tank and was protected against fire from ahead by an 8-mm armor plate. The tank was designed for long-range operations with the 115-liter auxiliary tank at a fuel mixture of 25 percent. Installation of the 300-liter auxiliary tank under the fuselage instead of the 115-liter tank meant that part of the cold-start mixture had to be dispensed with.

Equipment

Equipment was taken from the A-9 series. Changes included the hydraulic systems made necessary by the change to hydraulic undercarriage and landing flaps as well as minor modifications resulting from the installation of the Jumo 213 A.

Specific Equipment

Guns:
(see: "Weapons and Tank Arrangement" p. 31)
The TA 152 A could be equipped with the following weapons:
Normal: 5 weapons

2 MG 151 in fuselage with 150 rounds per gun
2 MG 151 in win roots with 175 rounds per gun
1 MK 108 engine-mounted cannon with 85 to 90 rounds or 1 MK 103 engine-mounted cannon with 75 to 80 rounds per gun
Additional: 2 weapons
2 MK 108 in outer wings with 55 rounds per gun or 2 MG 151 in outer wings with 140 rounds per gun or 2 MK 103 beneath the outer wings with 40 rounds per gun

Of the above weapons the four MG 151 fired through the propeller disc and were controlled electrically. The weapons installations listed under "additional" were identical in every detail to the installations used in current and planned series.

Gravity Weapons

The carriage beneath the fuselage of bombs up to 500 kg was planned for fighter-bomber operations. The bombs were mounted on a Type 503 rack installed in the wing center section in front of the forward spar. The bombs were supported by four aerodynamically-shaped, adjustable support arms beneath the fuselage. The carriage of gravity weapons and additional equipment (210-mm rockets etc.) was possible to the same extent as on the Fw 190 A series then in production.

Structural Strength

The Ta 152 strength manual (memo dated 25 May 1943) prescribed a wing load coefficient of +6.5 or –3.0 for the design takeoff weight of 4,400 kg. The required load coefficient was established for all components borrowed from the A series and the dimensions of the new wing center section and other altered components was taken into

Type	TA 152 A	TA 152 B	TA 152
Engine	Jumo 213 A	Jumo 213 E	DB 603 G
Takeoff power H.P. (kW)	1,750 (1,297)	2,050 (1,508)	1,900 (1,397)

1 Performance with light weapons installation: 1 MK 108 engine-mounted cannon plus 2 MG 151 in fuselage

Maximum speed (kph)	682	742	685
at altitude of (m)	7 000	10 750	8 300
Service ceiling (m)	11 200	12 900	12 000
Climb rate at sea level (m/sec) 13.9	14.8	13.2	

2 Performance with medium weapons installation: as 1 plus 2 MG 151 in wing roots

Maximum speed (kph)	678	734	676
at altitude of (m)	7 000	10 750	8 300
Service ceiling (m)	10 900	12 600	11 680
Climb rate at sea level (m/sec) 15.1	16.0	14.4	

3 Performance with heavy weapons installation: as 2 plus 2 MK 108 beneath outer wings

Maximum speed (kph)	671 (692)[7]	728 (735)[8]	670 (686)[9]
at altitude of (m)	7 000 (8 000)	10 750 (13 400)	8 300 (10 000)
Service ceiling (m)	10 500	12 300	11 200
Climb rate at sea level (m/sec) 13.9	14.8	13.2	

[1] The data were taken from the performance comparison sheets contained in Technical Specification No. 270.

[2] With use of GM 1, performance increase of 280 H.P.

[3] With use of GM 1, performance increase of 320 H.P.

[4] With use of GM 1, three performance settings possible (135 H.P., 275 H.P., 415 H.P.)

consideration. With standard armament and a normal takeoff weight of 4,620 kg, the Ta 152 B with Jumo 213 E revealed a load coefficient of 6.2.

The following load coefficients apply to the various power plant configurations:

Jumo 213 A standard engine with VS 9
propeller   6.7
Jumo 213 E standard engine with VS 9
propeller   6.5
DB 603 G standard engine with VDM
propeller   6.7

As this performance comparison shows, the performance of the Ta 152 with Jumo 213 A (C) was almost identical with the DB 603 G alternate engine. Only with the Jumo 213 E engine was a performance in excess of 700 kph possible.

In addition the Jumo 213 E possessed an increased maximum boost altitude of 10 750 meters as the result of its two-stage supercharger and three-speed gearing.

22

The Prototypes Fw 190 V 19, Fw 190 V 20 and Fw 190 V 21

A total of three prototypes were produced at Adelheide for the testing of the planned Focke-Wulf Ta 152 A series. All three were conversions of former Fw 190 A-0 production machines which had originally been planned as prototypes for the Fw 190 C-1 series, which was not proceeded with. As a rule, at this time the conversion of production aircraft into prototypes took place at Adelheide (Delmenhorst) near Bremen. The prototype aircraft were subsequently flown to the Focke-Wulf test center at Hannover-Langenhagen for performance trials. Testing at Langenhagen had only begun in 1943. The original test sites were Bremen and later Hamburg-Wenzendorf. In addition to the testing of new designs, Langenhagen was also the site of production and assembly of the new twin-engined, all-wood Ta 154, which was planned as a high-speed night fighter.

Construction of the two originally-planned prototypes for the TA 152 A-1 (TA 152 V 1, WNr. 250 001 and TA 152 V 2, WNr. 250 002) was canceled. Both prototypes were supposed to receive the new wing (19.5 m2) and an armament of one MK 108 engine-mounted cannon and MG 151 cannon above the engine and in the wing roots.

Forward view of the Fw 190 V 20. The heavy exhaust staining from the flame damper installation is visible aft of the new ejector-type engine cowling.

Fw 190 V 20, *Werknummer* 0 042, registration TI + IG, prototype for the Ta 152 A, now with flame dampers removed and conventional ejector exhausts. The aircraft made its first flight on 23 November 1943 with Hans Sander at the controls.

Summary of Flight Testing Results

In the initial phase of flight testing the Fw 190 V 19 there were considerable problems with rough running of the Jumo 213 A engine. The reason for this was resonance phenomenon of the airframe with the engine caused by the Jumo 213 A's high speed of 3,250 revolutions per minute.

No significant improvement in smoothness could be achieved until the installation of the Jumo 213 CV engine, which had a different firing sequence. During the course of flight testing it was also discovered that the propeller also exercised a strong influence on the running smoothness of the engine. For example, the Fw 190 V 20 with the Schwarz Company's VS 9 propeller experienced less vibration than

with the Heine Company's VS 9 which was installed on 4 February 1944. The reason was the unbalanced state of the propeller, to which the Jumo 213 reacted quite drastically.

During flight testing the Fw 190 V 19 made a crash landing as a result of failure of the right undercarriage leg locking bolt. The damage was repaired, however, and it was possible to resume testing.

The Fw 190 V 20 was the first aircraft to be fitted with the flame damper system. Focke-Wulf suspected that use of this system would result in a considerable loss of speed, which was confirmed during trials. With the flame damping installation the Fw 190 V 20 achieved a maximum speed of 657 kph at an altitude of 7,600 meters, or

Me 209 V 5, SP + LJ, made its first flight on 3 November 1943. It was the first prototype of the Me 209 A, a competitor of the Ta 152 A.

approximately 35 kph less than expected. During trials the Fw 190 V 21 exhibited excessively low coolant temperatures at low outside air temperatures. This problem was solved by modifying the radiator gill actuating rods, resulting in a significant improvement in the engine's winter operating performance.

The Fw 190 V 21 also encountered a problem with rough running of the engine, which like that of the V 20 was improved by installing a new propeller. The Fw 190 V 21 had a somewhat modified flame damper installation with an 85% straight cut exhaust gas collector pipe. In spite of this there was no improvement in maximum level speed compared to the Fw 190 V 20, which had a 100% straight cut exhaust gas collector pipe. The Fw 190 V 21 achieved a maximum speed of 540 kph at sea level and on 5 May 1944 was handed over to the Rechlin Test Station for trials. Since the Ta 152 A series did not come to fruition, the Fw 190 V 20 and V 21 were supposed to be converted into test beds for the new Daimler Benz DB 603 L engine for the planned Ta 152 C series. On 5 August 1944, however, the Fw 190 V 20 was heavily damaged in a bombing raid on Langenhagen and was not rebuilt for testing of the Ta 152 C series.

Rear view of the Fw 190 V 20 with cooling gills open.

25

List of Prototypes

1	**Fw 190 V 19**
Werknummer:	0 041
Registration:	— + —
First flight:	7 July 1943
Power Plant:	Junkers Jumo 213 A No. 100 152 082 Jumo 213 A No. 100 152 160 Jumo 213 A No. 100 1570 009
Purpose:	Investigation of handling qualities, engine performance flights and testing of the hydraulic system.
Remarks:	Initially retained the tail surfaces of the Fw 190 D series then converted to new tail of the Fw 190 C series, which was similar to that of the TA 152. Fuselage extension 500 mm, no armament, later installation and testing of MK 103 engine-mounted cannon. Conversion to flame damper installation planned but not carried out as a result of the crash of the V 19. Engine changed several times during the course of testing (see above).
Fate:	Crashed on 16 February 1944.
2	**Fw 190 V 20**
Werknummer:	0 042
Registration:	TI + IG
First Flight:	23 November 1943 (pilot Hans Sander)
Power Plant:	Junkers Jumo 213 CV No. 100 1570 010 standard engine
Purpose:	Testing of the flame damper installation, engine function checks, determination of horizontal speeds, effect of bulged engine cowling in flight, investigation of fuel supply system, static endurance runs, testing of the hydraulic system for undercarriage and landing flaps.
Remarks:	Fuselage extension 500mm, C tail, takeoff weight 3 900 kg, no armament, pressurized cockpit hood, flame damper installation was removed during testing. It was planned that the V 20 would become the V 20/U1 for testing of the new TA 152 C series (conversion to DB 603 L engine), conversion was not carried out after aircraft was heavily damaged in air raid on Langenhagen on 5 August 1944.
Fate:	Destroyed in bombing raid on Langenhagen on 5 August 1944.
3	**Fw 190 V 21**
Werknummer:	0 043
Registration:	TI + IH
First flight:	13 March 1944, pilot Bernhard Märschel
Power Plant:	Junkers Jumo 213 CV No. 100 1570 012 standard engine
Purpose:	Testing of flame damper installation, function checks, determination of horizontal speeds, handling quality checks, beginning of speed, boost pressure and fuel consumption trials at low altitude, testing of hydraulic system.
Remarks:	Fuselage extension 500 mm, C tail, takeoff weight 3 890 kg, no armament, wing area 19.5 m2, gun ports in engine cowling open, handed over to Rechlin Test Station on 5 May 1944. Was converted as Fw 190 V 21/U1 for testing of the TA 152 C series (DB 603 L engine installed), handed over to Daimler Benz for engine trials (DB 603 E) on 18 November 1944.

Summary

No serious shortcomings became apparent during the testing of the prototypes for the Ta 152 A series.

The flame damper installation originally planned for the series had such a negative effect on performance that it was officially dropped on 18 April 1944. But by that time the Ta 152 A was fully ready for production. The RLM's decision not to build the Ta 152 A in quantity was therefore all the more inexplicable, for the Focke-Wulf Ta 152 A's performance was clearly superior to that of the Fw 190 A with the BMW 801 radial engine at altitudes above 5,000 meters. The choice of power plants, especially the Jumo 213 E with three-gear transmission and two-stage supercharger then in development, suggested significant room for development. The Jumo 213 A (C), equipped with a single-stage supercharger and a two-gear transmission, had meanwhile achieved the necessary level of reliability and was successfully installed in the Fw 190 D-9 series from September 1944.

Top:
Right side view of the Fw 190 V 20 with the new supercharger air intake and flame damper installation with straight-cut collector pipe.

Center:
Left side view of the Fw 190 V 20 with detail view of the new flame damper installation.

Bottom:
View of the flame damper installation with engine cowling removed.

27

Fw 190 V 21, third prototype of the Ta 152 A, with flame damper installation, hydraulic undercarriage and full camouflage scheme.

Three quarter view of the Fw 190 V 21, WNr. 0 043, registration TI + IH. In contrast to the Fw 190 V 20, the V 21 had a straight-cut collector pipe.

		Focke-Wulf Flugzeugbau G.m.b.H. Abt: Flugmechanik L			**Horizontalgeschwindigkeit über der Flughöhe**						**Leistungsvergleich Ta 152**			

Kurve Nr.	Flugzeug Muster	Fluggewicht	Fläche	Motor	Motor-beanspruchung	Drehzahl	Ladedruck	Kraft-stoff	Bemerkungen			
-	-	kg	m²	-	-	U/min	ata	-	Motor	Rumpf	Flügel innen	Flügel außen
1	Ta 152	4600	19,5	Jumo 213 A-1	Start- und	3200	-	B4	1×Mk 108 +2×MG 151 + 2×MG 151 +2×Mk 108 (1×60 Schuß) (2×175) (2×150) (2×60)			
2	Ta 152	4635	"	DB 603 G	Not-	2700	1,5/1,4	C3	"	"	"	"
3	Ta 152	4700	"	Jumo 213 E	leistung	3200	-	C3	"	"	"	"

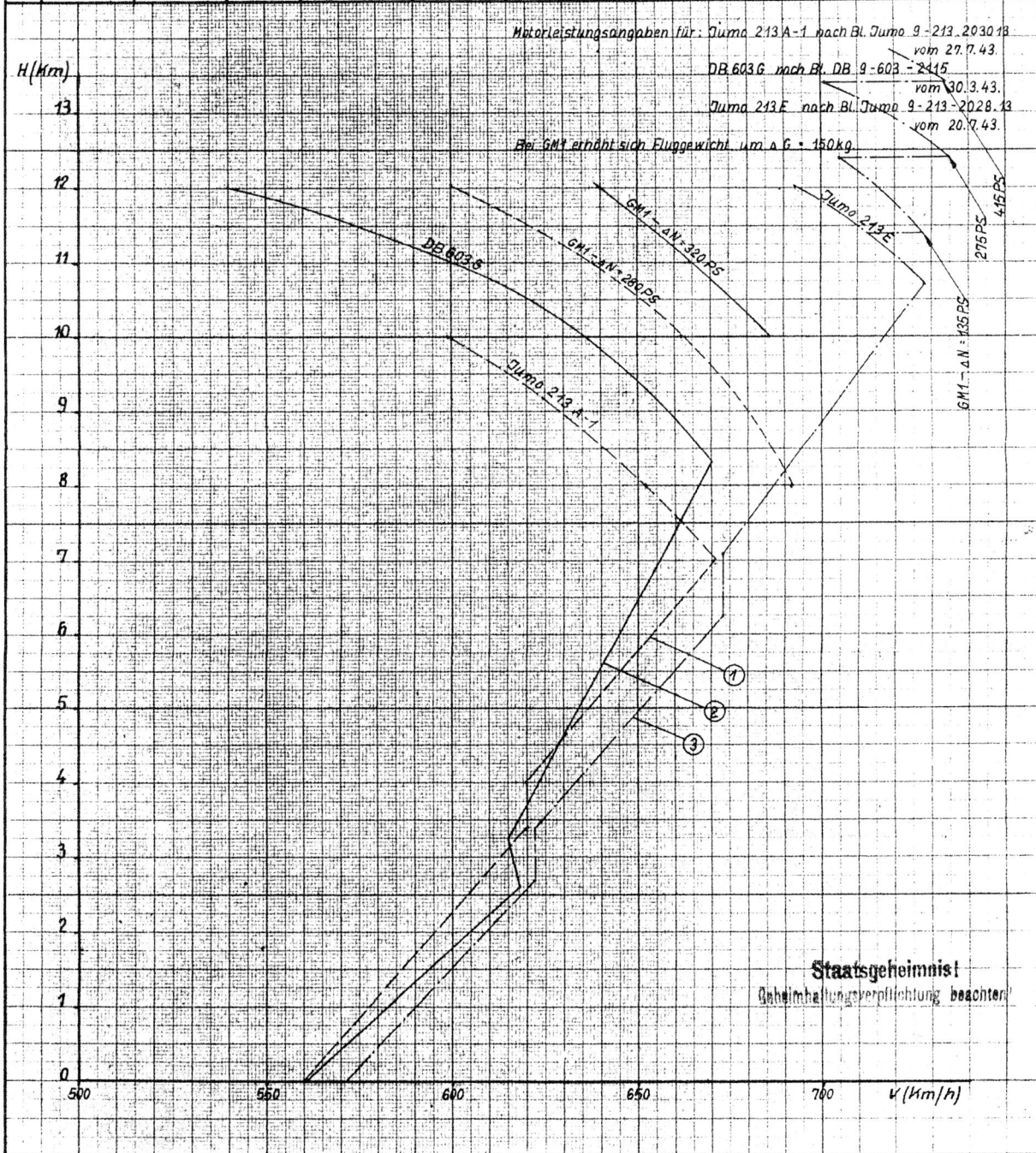

Motorleistungsangaben für: Jumo 213 A-1 nach Bl. Jumo 9-213-2030 13 vom 27.7.43.
DB 603 G nach Bl. DB 9-603-2115 vom 30.3.43.
Jumo 213 E nach Bl. Jumo 9-213-2028.13 vom 20.7.43.
Bei GM1 erhöht sich Fluggewicht um ΔG = 150 kg.

DB 603 G
GM1-ΔN=320 PS
GM1-ΔN=260 PS
Jumo 213 A-1
Jumo 213 E
GM1-ΔN=135 PS
275 PS
415 PS

n J. 152.000-001

Datum: 4.9.43.

Bearbeiter: Jansen

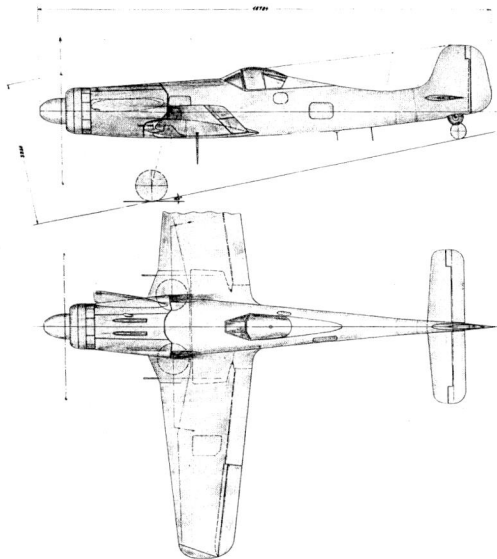

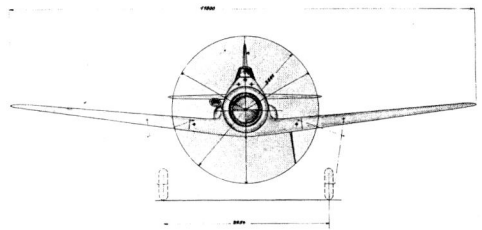

Jagdflugzeug Ta 152 A (Übersichtszeichnung)

mit Jumo 213 A Einheitstriebwerk,
und Fla.V. Anlage

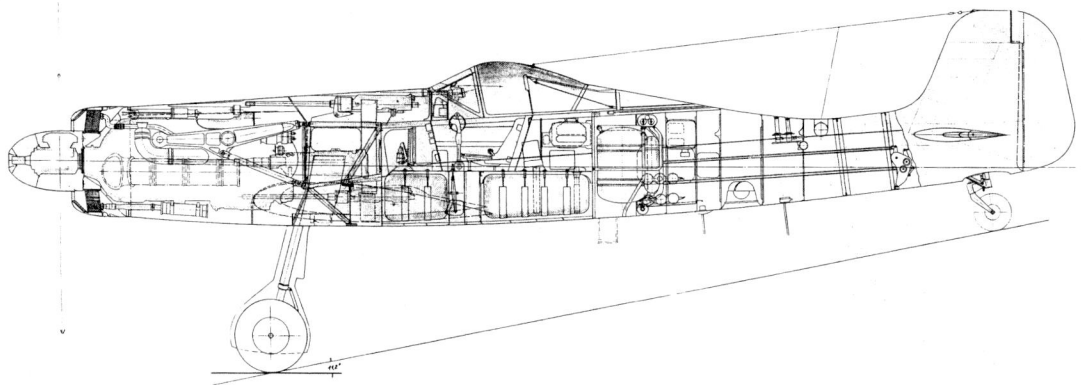

Jagdflugzeug Ta 152 A (Längsschnitt)

mit Jumo 213 A ETW und GM1-Anlage,
und Fla.V. Anlage

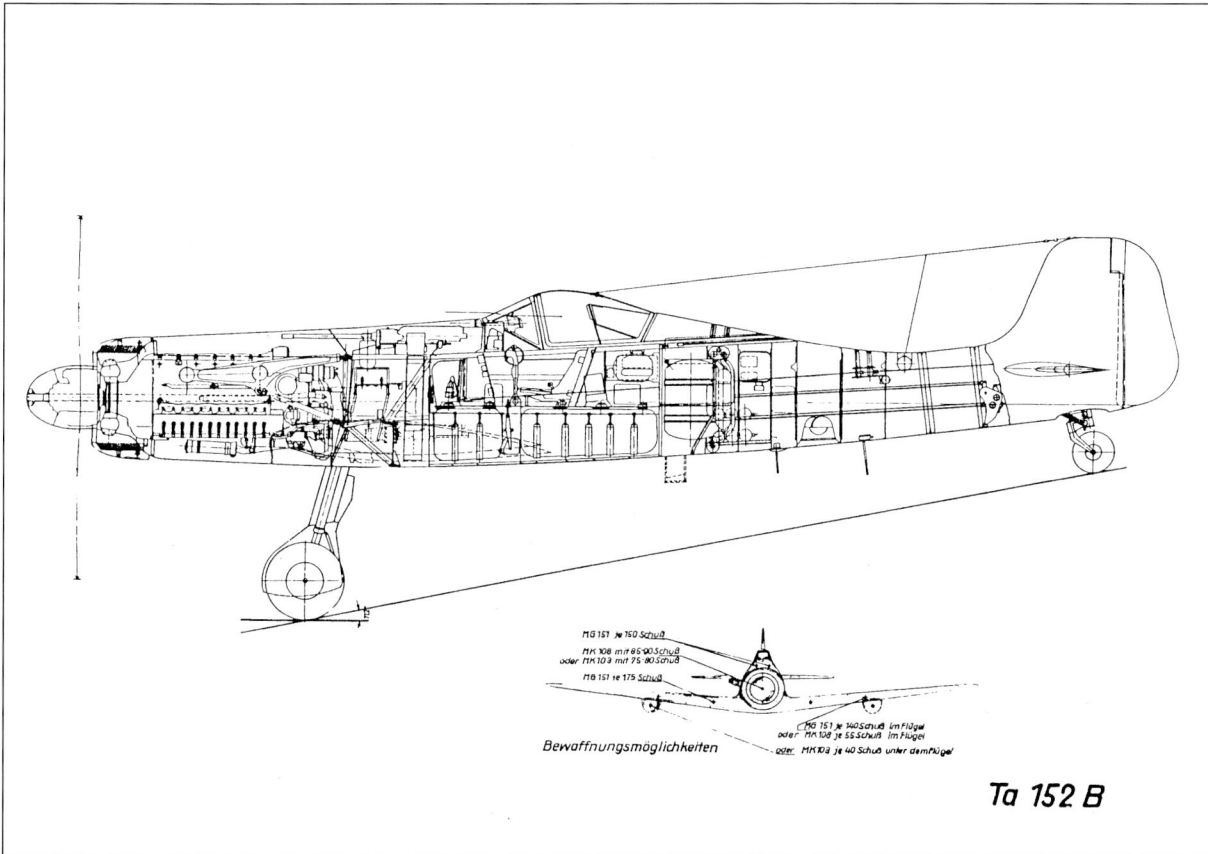

MG 151 je 150 Schuß
MK 108 mit 85-90 Schuß
oder MK 103 mit 75-80 Schuß
MG 151 je 175 Schuß

MG 151 je 140 Schuß im Flügel
oder MK 108 je 55 Schuß im Flügel
oder MK 103 je 40 Schuß unter dem Flügel

Bewaffnungsmöglichkeiten

Ta 152 B

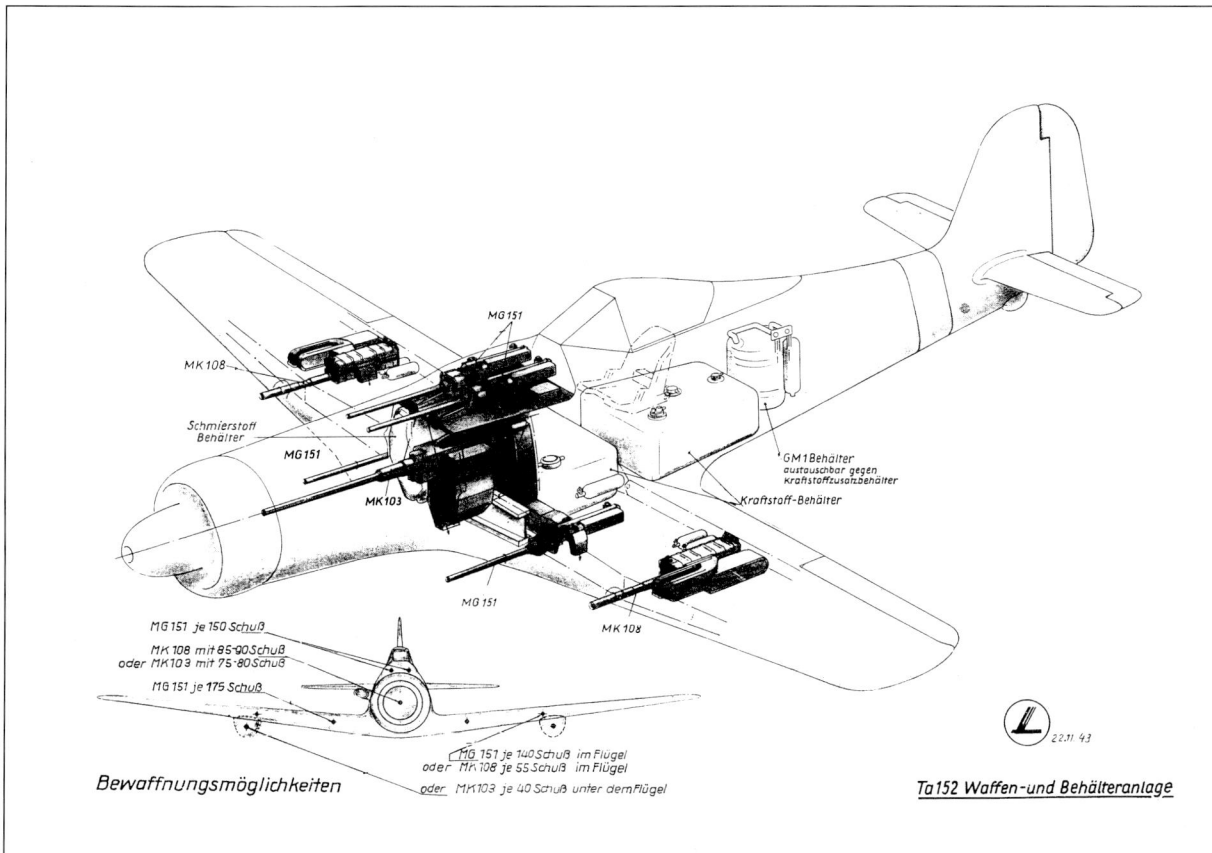

MG 151
MK 108
Schmierstoff Behälter
MG 151
MK 103
MG 151
MK 108

GM 1 Behälter
austauschbar gegen
Kraftstoffzusatzbehälter
Kraftstoff-Behälter

MG 151 je 150 Schuß
MK 108 mit 85-90 Schuß
oder MK 103 mit 75-80 Schuß
MG 151 je 175 Schuß

MG 151 je 140 Schuß im Flügel
oder MK 108 je 55 Schuß im Flügel
oder MK 103 je 40 Schuß unter dem Flügel

Bewaffnungsmöglichkeiten

22.11.43

Ta 152 Waffen- und Behälteranlage

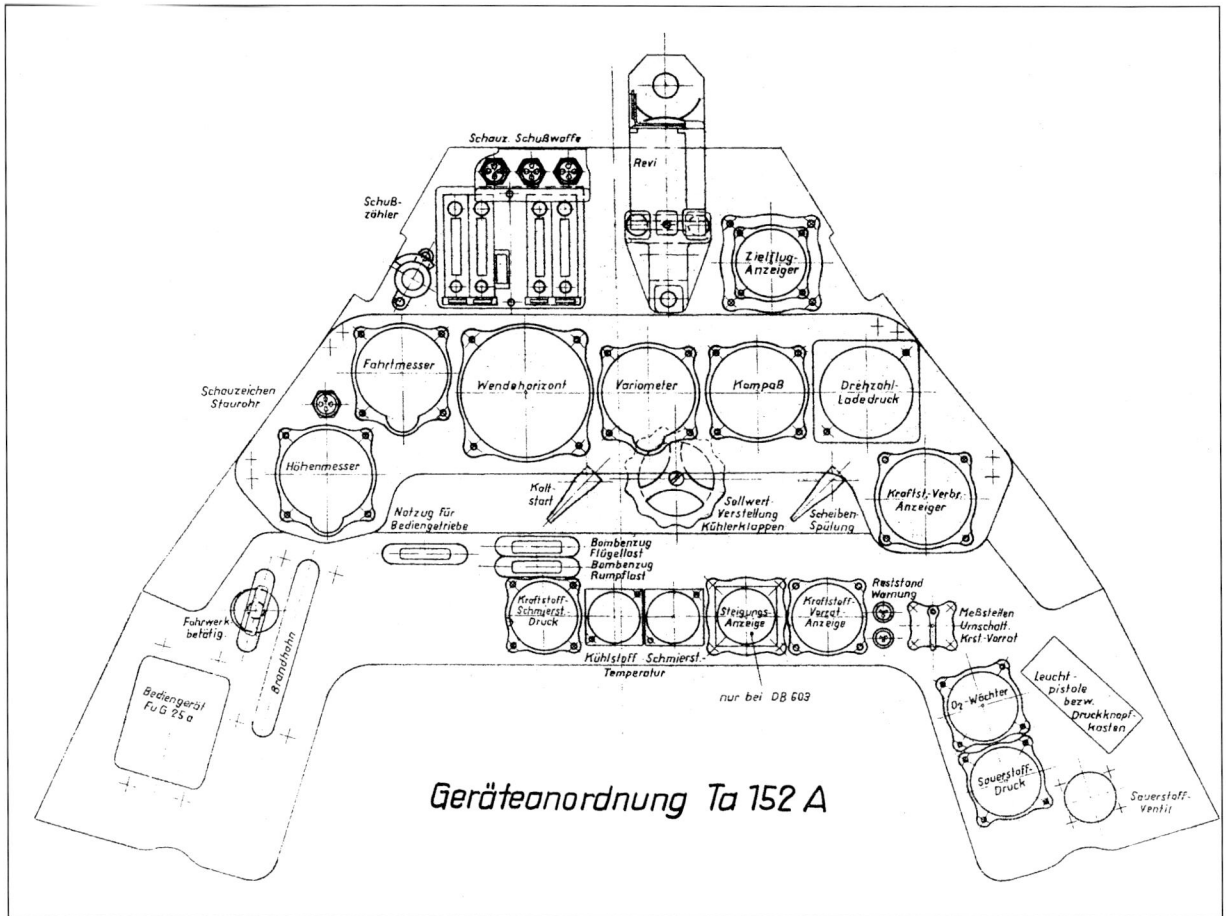

Geräteanordnung Ta 152 A

Schauz. Schußwaffe
Schuß-zähler
Revi
Zielflug-Anzeiger
Schauzeichen Staurohr
Fahrtmesser
Wendehorizont
Variometer
Kompaß
Drehzahl-Ladedruck
Höhenmesser
Kalt-start
Sollwert-Verstellung Kühlerklappen
Scheiben-Spülung
Kraftst.-Verbr. Anzeiger
Notzug für Bediengetriebe
Bombenzug Flügellast Bombenzug Rumpflast
Reststand Warnung
Fahrwerk betätig.
Kraftstoff-Schmierst.-Druck
Steigungs-Anzeige
Kraftstoff-Vorrat-Anzeige
Meßstellen-Umschalt. Krst-Vorrat
Brandhahn
Kühlstoff Schmierst.-Temperatur
nur bei DB 603
O₂-Wächter
Leucht-pistole bezw. Druckknopf-kasten
Bediengerät Fu G 2S a
Sauerstoff-Druck
Sauerstoff-Ventil

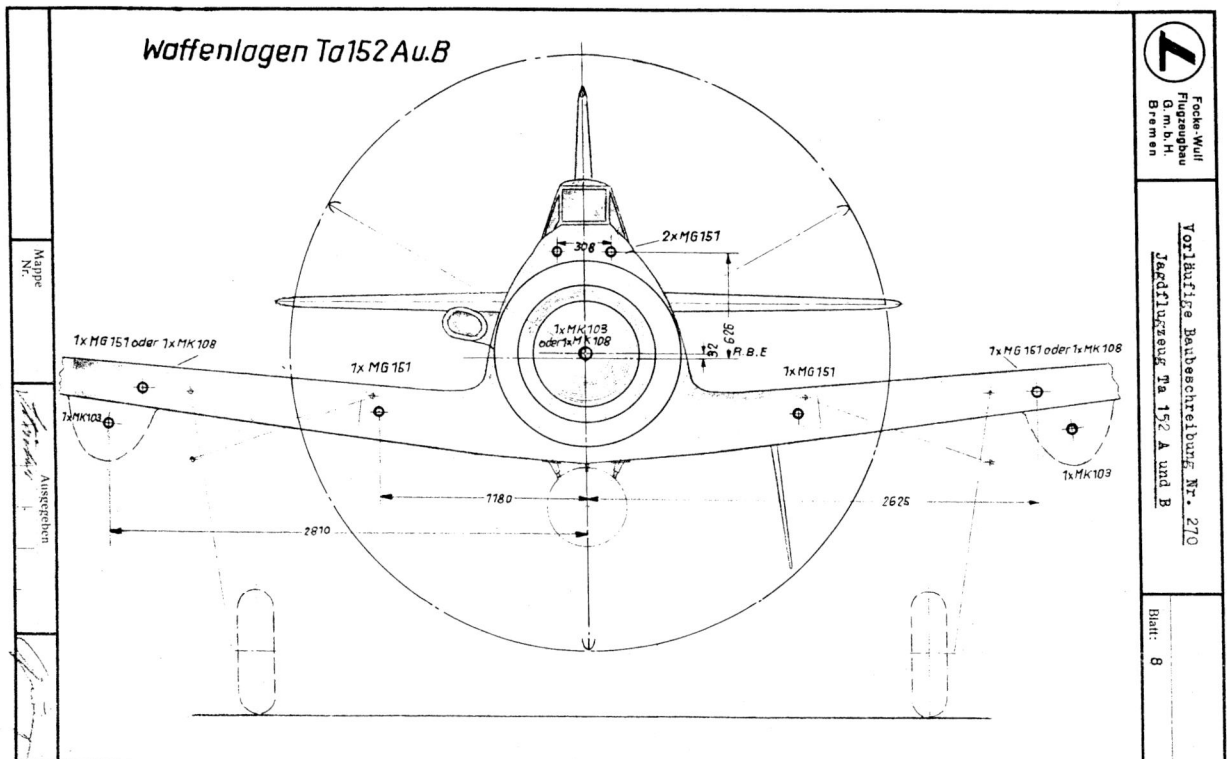

Focke-Wulf Flugzeugbau G. m. b. H.   Nr. 26 a

Waffenlagen Ta 152 A u. B

2 x MG 151
30 B
1 x MK 108 oder 1 x MK 108
B.O.B
R.B.E
1 x MG 151 oder 1 x MK 108
1 x MG 151
1 x MG 151
1 x MG 151 oder 1 x MK 108
1 x MK 103
1 x MK 103
1180
2625
2810

Mappe Nr.
Ausgegeben

Focke-Wulf Flugzeugbau G. m. b. H. Bremen

Vorläufige Baubeschreibung, Nr. 270
Jagdflugzeug Ta 152 A und B

Blatt: 8

32

Focke-Wulf
Flugzeugbau
G. m. b. H.
Bremen

Vorläufige Baubeschreibung Nr.270
Jagdflugzeug Ta 152 A u. B

Blatt: 7

Gewichtsaufstellung

Bezeichnung	Triebwerk		
	Jumo 213 A m.Fla-V-Anlage	Jumo 213 E o.Fla-V-Anlage	DB 603G
Rumpfwerk m. Panzer f. Schmierst.	360,0	360,0	360,0
Fahrwerk	295,0	295,0	295,0
Leitwerk	127,0	127,0	127,0
Steuerwerk	31,0	31,0	31,0
Tragwerk	509,0	509,0	509,0
Triebwerk (vergröß. Behälter)	1805,0	1909,0	1832,0
Ausrüstung normal	193,0	193,0	193,0
Zweckausrüstung	384,0	384,0	384,0
Ballast	–	30,0	–
Rüstgewicht	3704,0	3838,0	3731,0
1 Führer	100,0	100,0	100,0
Kraftstoff 595 l (B4 bzw. C3)	440,0	464,0	464,0
Schmierstoff	40,0	40,0	40,0
Munition 2 MG 151 im Rumpf je 150 Sch.	67,0	67,0	67,0
" 2 MG 151 i.Flügelw." 175 Sch.	78,0	78,0	78,0
" 1 MK 108 im Motor 60 Sch.	36,0	36,0	36,0
Zuladung	761,0	785,0	785,0
Normal-Fluggewicht 1	4465,0	4623,0	4516,0
Normal-Fluggewicht	4465,0	4623,0	4516,0
GM1-Anlage	38,0	38,0	38,0
GM1-Füllung 85 l	102,0	102,0	102,0
Ausbau von Ballast	–	–15,0	–
Normal-Fluggewicht 2 mit GM1	4605,0	4748,0	4656,0
Normal-Fluggewicht 1	4465,0	4623,0	4516,0
je Seite 1 MK 108 im Flügel außen	175,0	175,0	175,0
Munition 2 MK 108 je 55 Schuß	66,0	66,0	66,0
Fluggewicht 3 mit schwerer Bewaffnung ohne GM1-Anlage	4706,0	4864,0	4757,0
Fluggewicht 3	4706,0	4864,0	4757,0
GM1-Anlage	38,0	38,0	38,0
GM1-Füllung 85 l	102,0	102,0	102,0
Fluggewicht 4 mit schwerer Bewaffnung mit GM1-Anlage	4846,0	5004,0	4897,0

Bei Einbau der Motorkanone MK 103 statt MK 108 bei gleicher Munitionsmenge
(60 Schuß) erhöhen sich obige Fluggewichte um 105 kg.
Die den Leistungsrechnungen zugrundegelegten Gewichte unterscheiden sich
nur unwesentlich von obigen Angaben. Die Mehrgewichte sind auf die ver-
größerte Kraftstoffanlage und diverse kleine Änderungen zurückzuführen.

15.12.43 Stä/Bu.   Mappe Nr.   Ausgegeben

Versuchsmuster Fw 190 V 20, Werknr.: 0042, Kennung TI + IG, für Ta 152 A - Serie

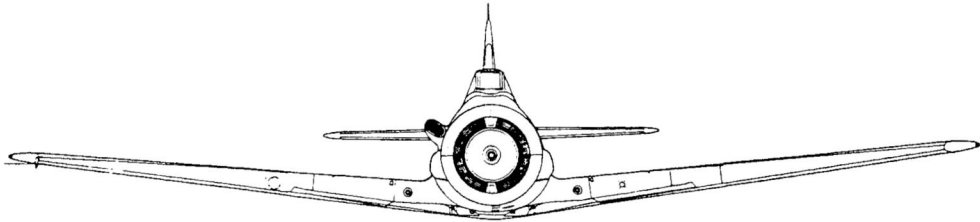

Vorderansicht der Fw 190 V 20

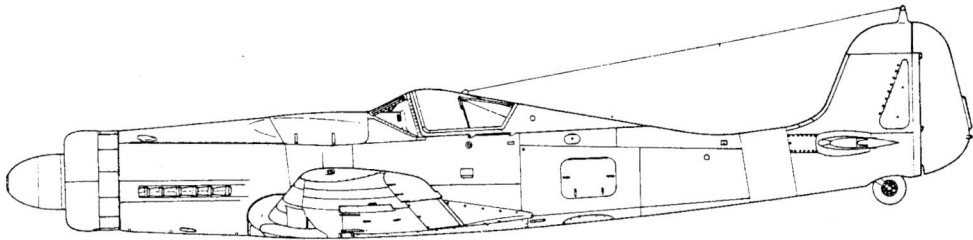

Seitenansicht nach Ausbau der Flammenvernichteranlage

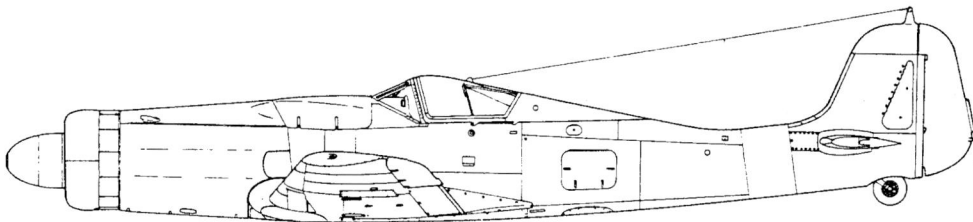

Seitenansicht mit Flammenvernichteranlage

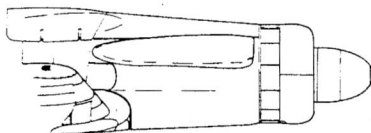

Langer Laderlufteinlaß mit FlaVAnlage   Langer Laderlufteinlaß mit normaler Schubstrahlanlage

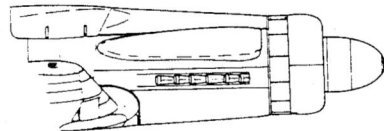

The Heavy Fighter – The Ta 152 B-5

In autumn 1944 the RLM conducted a strict rationalization of aircraft types which meant the end of the Me 410, a classic *Zerstörer* (heavy fighter). As a result the potential roles of the Ta 152 came to include that of the heavy fighter, which in the past had always been a twin-engined machine. In order to meet the RLM's requirements Focke-Wulf developed the Ta 152 B-5 *Zerstörer*. The Ta 152 B-5 used the same airframe as the Ta 152 C-3 but with the following changes:

a   Jumo 213 E power plant
b   Armament of one MK 103 engine-mounted cannon and two MK 103 in the wing roots
c   Possible use of MW 50 injection.

From the beginning it was intended that the Ta 152 B-5 would be manufactured with bad weather equipment as the Ta 152 B-5/R11. In January 1945 it was originally planned that production would begin at Erla in May 1945 and Gotha in July 1945,[1] however two months later the question of pro-

TA 152 B-5 Specification

Purpose:	Single-seat heavy fighter
Configuration:	Single-engine, low-wing cantilever monoplane with hydraulically retractable undercarriage
Power Plant:	Jumo 213 E with methanol-water injection (MW 50)
Dimensions:	Wing area 19.5 m2 Wingspan 11.0 m Aspect ratio 6.2 Vertical tail area 1.77 m2 Horizontal tail area 2.82 m² Maximum length 10.8 m Maximum height 3.38 m
Normal takeoff weight:	TA 152 B-5 heavy fighter mission 5,450 kg
Armament:	TA 152 B-5 2 MK 103 in wing roots with 44 rounds per gun 1 MK 103 engine-mounted cannon with 80 rounds
Armor:	Engine armor 10/6 mm 62 kg Cockpit armor 20, 15, 10, 8, 5 mm 88 kg Armored windscreen 70 mm
Equipment:	FuG 16 ZY, FuG 25a, FuG 125, K 23 autopilot, Revi 16b
Fuel Capacity:	TA 152 B-5 Normal in fuselage: 594 l B4 Additional 6 wing tanks 470 l B4 Drop tank under fuselage 300 l B4
Performance:	Maximum speed: 529 kph at sea level at emergency power (5 min.) Maximum speed: 683 kph at 10,700 m at emergency power (5 min) Maximum speed: 710 kph at 9,500 m Maximum ceiling using MW 50: 11,600 m
Range:	1,165 km without 300-l drop tank

[1] Focke-Wulf production type overview 3/1/1945.

Side view of the Fw 190 V 68, DU + JC, Werknummer 170 003, MK 103 weapons test-bed for the Ta 152 B.

duction was again thrown completely open on account of the war situation.[2]

In order to test the wing-mounted MK 103 cannon, the Fw 190 V 53, *Werknummer* 170 003, registration DU + JC, which had originally served to test the Fw 190 D-9, was converted into an MK 103 test bed as the Fw 190 V 68. The Fw 190 V 68 was not ready to begin testing until 13 December 1944, and at the end of 1944 it was delivered to the Tarnewitz Testing Station for weapons trials.

Three further prototypes built at Sorau were supposed to take part in the preliminary testing of the Ta 152 B-5: Ta 152 V 19 (WNr. 110 019), Ta 152 V 20 (WNr. 110 020), and Ta 152 V 21 built to Ta 152 B-5/R11 standard. Although it was anticipated that the V 19 and V 20 would be ready to fly in March 1945 and the V 21 in April, no maiden flights were made by these prototypes by 13 March 1945. Thus the only prototype to be tested was the V 68, which was fitted with wheel well doors.

[2] Focke-Wulf production type overview 21/3/1945.

Detail view of the MK 103 in the wing root, wheel well cover clearly visible.

MK 103 fairing on the right wing, undercarriage position indicator rod clearly visible.

Ta 152 B-5, MK 103 installed in right wing.

Development of the Fighter-Bomber
The Focke-Wulf Ta 152 C

The development and design of the Ta 152 C was essentially based on the following requirements issued by the RLM:

1   For increased performance, installation of the DB 603 LA or DB 603 L in the Fw 190 A series airframe with the minimum possible changes and maximum possible use of existing jigs and tools, ensuring that installation of the Jumo 213 E would be possible.

2   Strengthening of the central armament, in particular installation of an MK 103 or MK 108 engine-mounted cannon. The requirement for minimum possible modifications also applied here.

3   Installation of larger 740x210 wheels to cope with the increased takeoff weight resulting from the above changes.

4   Incorporation of a wing tank system for increased range without compromising the external aerodynamics.

By the beginning of 1944 it was already apparent that there was little development potential left in the Fw 190 powered by the BMW 801 radial engine. The performance differential became even more marked with the appearance over the German Reich of the latest high-performance Allied fighter aircraft (P-47D Thunderbolt, P-51D Mustang, Spitfire XIV and Tempest V). While these Allied fighters were all capable of maximum speeds of around 700 kph, the Fw 190's maximum speed was reduced by additional armor and increased weight. It was therefore obvious that the only way to close the gap was with a higher performance version in the form of the Ta 152.

Fw 190 V 21/U1, engine test-bed for the Ta 152 C, after conversion to the DB 603 E.

Technical Specification No. 290 for the Ta 152 C dated 5 January 1945

Ta 152 C-1 and C-3 Specification Sheet

Type:	Single-seat fighter aircraft (fighter-bomber)
Configuration:	Single-engine, low-wing cantilever monoplane with hydraulically retractable undercarriage.
Structural strength:	Load coefficient of 6.3 at medium fighter mission weight of 5 000 kg
Power Plant:	DB 603 LA with methanol-water (MW 50) Later DB 603 L with supercharger intercooler

Dimensions:

Wing area:	19.5 m²
Wingspan:	11.0 m
Aspect ratio:	6.2
Vertical tail area:	1.77 m2
Horizontal tail area:	2.82 m2
Maximum length:	10.80 m
Maximum height:	3.38 m
(at right angle to propeller)	
Mainwheels:	740x210 mm
Tailwheel:	380x150 mm

Normal takeoff weight:

Fighter mission TA 152 C-1	5,300 kg
Fighter-bomber mission TA 152 C-1	5 500 kg

Armament:

TA 152 C-1
2 MG 151/20 in upper fuselage with 150 rounds per gun
2 MG 151/20 in wing roots with 175 rounds per gun
1 MK 108 engine-mounted cannon with 90 rounds

TA 152 C-3
2 MG 151/20 in upper fuselage with 150 rounds per gun
2 MG 151/20 in wing roots with 175 rounds per gun
1 MK 103 engine-mounted cannon with 80 rounds

Armor:

Engine armor 10/6 mm	62 kg
Cockpit armor 20, 15, 10, 8, 5 mm	88 kg
Armored windscreen 70 mm	
Total weight of armor	150 kg

Equipment: FuG 16 ZY, FuG 25a, FuG 125, K 23 autopilot, Revi 16b

Fuel Tanks:

TA 152 C-1, C-3
Normal in fuselage: 594 l B4, 140 l MW 50
Additional 6 wing tanks 470 l B4
Drop tank under fuselage 300 l B4

The standard version originally planned for this role, the Ta 152 B powered by a Jumo 213 E, did not come to fruition at first on account of a change in plans by the RLM. Therefore the version now designated Ta 152 C was to be built with the DB 603 L without the supercharger intercooler but with MW 50 injection, however a change to the Jumo 213 E at a later date was to remain a possibility.

In contrast to the planned Ta 152 B, the Ta 152 C was to have no outboard wing armament, instead from the first aircraft it would have unprotected bag tanks in the wings. Other changes included new control linkages and propeller pitch control mechanism and a new MW 50 system.

Another interesting development was the design of the Ta 152 C as a torpedo carrier. This development work resulted in the Ta 152 C-1/R14, but the torpedo installation proceeded no further than the mockup stage.

Ta 152 V 6, *Werk-nummer* 110 006, VH + EY, was the first prototype of the Ta 152 C.

Detail view of the front end of the Ta 152 C V 7 with heavy exhaust staining on the engine cowling.

Fuselage

The following significant changes were made to the fuselage compared to that of the Fw 190 A-8:

The increased space required for the MK 103 engine-mounted cannon and the fuselage-mounted MG 151s made it necessary to lengthen the forward fuselage by 772 mm. In order to limit the procurement of new jigs and tools to an absolute minimum, the fuselage extension was bolted onto the existing engine attachment points. The wing, which had to be moved forward 420 mm for center of gravity reasons, was attached in the center of the extension section. The relocation of the wing also made it necessary to move the rear spar attachments and the corresponding fuselage bulkheads. The resulting change in the mounting of the forward tank made it necessary to redesign the tank compartment cover and the side skinning in the affected area.

In order to avoid having to accept a reduction in stability, especially directional stability as a result of the lengthened engine compartment, the aft fuselage was lengthened by inserting a 0.5-meter-long cylindrical section. The latter also served to accommodate the oxygen bottles and compressed

40

Front view of the Daimler Benz DB 603 L, which was planned for the production Ta 152 C.

Rear view of the new Daimler Benz DB 603 L engine, which was also supposed to be installed in the Dornier Do 335.

air bottles for the engine-mounted cannon which had been moved aft for center of gravity reasons.

Undercarriage

The undercarriage leg including shock strut and mounting was retained. The former electric drive was changed to a hydraulic system. Larger 740 x 210 wheels were installed on account of the aircraft's increased takeoff weight.

Control Surfaces

The ailerons, rudder, elevators and horizontal stabilizer were retained. The landing flaps were retained with minor modifications resulting from relocation of the actuators. The vertical fin was enlarged to 1.77 m^2. At the same time a strengthened tail-wheel assembly was fitted with tire measurements of 380x150 mm.

Daimler Benz DB 603 E engine, which was supposed to be installed in initial production Ta 152 Cs.

Flight Control System

The control system was largely retained, however the revised position of the wing and the fuselage extension resulted in some changes in the control linkages.

Wing Assembly

In order to provide propeller clearance from the larger wheels, it was necessary to move the latter outboard by 250 mm. This was made possible by increasing the span of the existing wing from 10.5 to 11 meters, which was achieved by inserting a 0.5-meter spar section in the center of the wing. In turn this made it necessary to redesign the wing-fuselage junction. As a result of the greater wingspan and the increase in the product of n x G, it became necessary to strengthen the skinning in the area of the inner wing. In order to facilitate repairs, it was planned to replace the former one-piece wing with a two-piece structure with a separation point on the aircraft centerline. The separation point consisted of wedge-shaped butt straps which were bolted to the top and bottom flanges of the forward spar.

Power Plant

Initial production machines were to be powered by the DB 603 E; however, the DB 603 LA, which had a considerably better high-altitude performance, was planned for

the Ta 152 C-1 and C-3. Methanol-water injection (internal cooling) was used to deal with the increased temperature of the supercharged air produced by the supercharger. The consumption of the methanol-water mixture (MW 50) was 90 liters per hour at climb and combat power and 190 liters per hour at takeoff and emergency power. For thermal reasons, takeoff and emergency power could only be used for three periods of ten minutes. The DB 603 LA was later replaced by the DB 603 L, which was equipped with a supercharger cooler instead of MW 50 internal cooling. The supercharger air cooler acted as a heat exchanger, bleeding its heat through a segment of the annular radiator.

Pertinent Data on the DB 603 LA Power Plant

Engine type:

- Liquid-cooled, twelve-cylinder, inline engine with two-stage supercharger: gearing 1 : 1.93

Detail view of the new engine cowling and supercharger air intake of the Ta 152 C.

Detail view of the exhausts and MG 151/20 wing cannon of the Ta 152 V 6.

- Liquid cooling: drum radiator with radial flow-through, radiator frontal area 58 dm².

- Lubricant cooling: nose radiator with axial flow-through, radiator frontal area 9 dm².

- Fuel: 87-octane B4, later 100-octane C3.

Propeller: three-bladed variable-pitch propeller with VDM hub and wooden blades. Propeller diameter 3.60 m. Blade width 12.2%.

Fuel System

It was possible to retain the 233-liter forward fuel tank from the Fw 190 A-8 series unchanged, however it was moved forward in keeping with the revised wing position.

As a result it was possible to increase the capacity of the rear tank by 70 liters to a total of 362 liters. This increased standard tankage to a total of 595 liters. The tanks were protected tanks with the following wall thicknesses: sides and bottom 16 mm, top 12 mm.

For increased range the following auxiliary tank installations were planned:

Ta 152 C-0

A 300-liter drop tank which was attached to an aerodynamically faired external mount, the so-called "TA 152 Tank Carriage," by grommets. Fuel transfer by supercharger air.

Ta 152 C-1, C-3

Six unprotected bag tanks in the wing with a total capacity of 470 l. Fuel transfer by supercharger air. The tanks were installed through access ports in the underside of the wing. For longer-range missions the 300-liter drop tank (eventually a 600-liter tank) described above could be used.

Lubrication System

The oil tank, with a total capacity of 72 liters, was installed on the right side of the forward fuselage extension next to the engine cannon. It was largely protected against fire from ahead by the engine. The tank's capacity of 61 liters was sufficient for operations with the fuselage tankage of 594 liters and the 300-liter auxiliary tank with a cold start mixture of 25 percent. For missions with greater tankage the cold start mixture had to be reduced accordingly.

Standard Equipment

Equipment was largely taken from the Fw 190 A-8 series. Exceptions included the hydraulic systems made necessary by the change to hydraulic undercarriage and landing flaps as well as minor modifications resulting from the installation of the DB 603 LA.

The most important planned equipment included: FuG 16 ZY radio system (transmit and receive), FuG 25a IFF set, FuG 125 radio guidance set, remote reading

Pertinent Data on the Prototypes of the Ta 152 C-1 Series

Fw 190 V 21/U1

Werknummer:	0 043
Registration:	TI + IH
First flight:	3 November 1944, pilot Märschel
Power plant:	Daimler Benz DB 603 E, serial no. 525 (B1 engine V 17)
Purpose:	Engine test-bed for the Ta 152 C-1 series
Remarks:	On 3 November 1944 pilot Märschel ferried the Fw 190 V 21/U1 from Adelheide to Langenhagen. On 18 November 1944 the V 21/U1 was handed over from Daimler Benz to Focke-Wulf at Langenhagen and on 19 November was ferried by Fw. Albrecht to Echterdingen (Daimler Benz). Daimler Benz began conversion to the DB 603 LA V 16. First flight by the V 21 with this engine on 10 December 1944.
	Flight testing of the V 21/U 1 by Daimler Benz with the DB 603 LA is known to have continued until March 1945

Ta 152 V 6

Werknummer:	110 006
Registration:	VH + EY
First flight:	12 December 1944, pilot Märschel
Power plant:	Daimler Benz DB 603 EC serial no. 0130 0145 (B1 engine V 19)
Purpose:	General function checks, testing of the hydraulic system, horizontal flights at combat and emergency power, climbing flights at combat power to determine rate of climb
Remarks:	Wing area 19.5 m², MW 50 system installed, heated canopy side panels, wooden landing flaps, C tail with 500-mm fuselage extension, 35 kg ballast in fuselage, later integral engine cowling. Takeoff weight: 4 370 kg, ETC 503 in fuselage.
	The V 6 was ready to fly on 6 December 1944. On 17 December Märschel ferried the V 6 from Adelheide to Langenhagen (see flight report).
Armament:	2 MG 151 in fuselage and 2 MG 151 in wing roots.

Ta 152 V 7

Werknummer:	110 007
Registration:	CI + XM
First flight:	8 January 1945, pilot Märschel
Power plant:	Daimler Benz DB 603 engine cowling, serial no. 0130 0147 (B1 engine V 20)
Purpose:	General function checks, testing of the hydraulic system, horizontal flights at combat and emergency power. Climbing flights at combat power to determine rate of climb.
Remarks:	Wing area 19.5 m². MW 50 system installed. Integral engine cowling, heated side panels, wooden landing flaps, C tail with 500-mm fuselage extension. Ballast 9.2 kg.
	The V 7 was ready to fly on 5 January 1945. Ferried from Adelheide to Langenhagen by pilot Märschel on 16 January. Hans Sander test flew the V 7 at Langenhagen on 27 January, 3 February and 6 February 1945. Conversion from DB 603 EC to DB 603 LA did not take place at Langenhagen until March 1945.
Armament:	2 MG 151 in fuselage and 2 MG 151 in wing roots.

Ta 152 V 8

Werknummer:	110 008
Registration:	GW + QA
First flight:	15 January 1945, pilot Märschel
Power plant:	Daimler Benz DB 603 EC serial no. 0130 0150 (B1 engine V 21)
Purpose:	General function checks, testing of the hydraulic system, horizontal flights at combat and emergency power. Climbing flights at combat power to determine the rate of climb.
Remarks:	Wing area 19.5 m², MW 50 system installed, integral engine cowling, heated side panels, wooden landing flaps, C tail with 500-mm fuselage extension. The V 8 was ready to fly on 14 January 1945. Bernhard Märschel ferried the V 8 from Adelheide to Langenhagen on 20 January 1945. Testing of the Ta 152 V 8 by the Rechlin Testing Station began in February 1945.
Armament:	2 MG 151 in fuselage, 2 MG 151 in wing roots.

Rear side view of the
Ta 152 C in the winter
of 1945.

compass, turn and bank indicator and the usual monitoring and navigation equipment, LGW K 23 automatic pilot.

Specific Equipment
Guns:

Ta 152 C-0, C-1: 2 MG 151/20 (20-mm) in upper fuselage with 150 rounds per gun, 2 MG 151/20 in wing roots with 175 rounds per gun, 1 MK 108 engine-mounted cannon (30-mm) with 90 rounds.

Ta 152 C-3: 2 MG 151/15 (15-mm) in upper fuselage with 150 rounds per gun, 2 MG 151/15 in wing roots with 175 rounds per gun, 1 MK 103 engine-mounted cannon (30-mm) with 80 rounds.

Additional: 2 weapons

Of the above weapons the four MG 151 fired through the propeller disc and were controlled electrically. Various types of rocket projectile could be mounted beneath the wings.

Gravity Weapons
The carriage beneath the fuselage of bombs up to 500 kg was planned for fighter-bomber operations. The bombs were mounted on a Type 503 rack installed in the wing center section in front of the forward spar. The bombs were supported by four aerodynamically-shaped, adjustable support arms beneath the fuselage. When an external stores rack was installed the Ta 152 tank carriage described under "Fuel System" was used.

Passive Protection
Cockpit armor was expanded and strengthened in keeping with the increased demands resulting from the heavier armament carried by Allied machines. Further strengthening, in particular the back armor to 15 mm, was planned.

Ta 152 C-1/C-3 Armor

	Armor thickness (mm)	Armor Weight (kg)
Forward ring armor (engine)	15	31.5
Rear ring armor (engine)	8	30.0
Armor in front of windscreen	15	14.0
Armored windscreen	70	22.5
Back armor	8	18.2
Shoulder protection	5	5.9
Armor plate on bulkhead	55	7.9
Head armor	20	20.0
Total weight of armor		150.0

Front view of the Ta 152 V 7, CI + XM.

Structural Strength

A load coefficient of 6.3 was calculated for the medium fighter weight of 5,000 kg. The certain negative load coefficient was –3.0. For the fighter-bomber mission with a 500-kg bomb and a weight of 5,500 kg the load coefficient was 5.6. Engine bearer load coefficient was 6.8.

The Prototypes Fw 190 V 21/U1, Ta 152 V 6, Ta 152 V 7, Ta 152 V 8

In the beginning no less than seventeen prototypes were planned for the new Ta 152 C series.[1] Of these sixteen were to be constructed as completely new aircraft in the Sorau factory. Another test-bed was to be created by converting the Fw 190 V 21, which would become the Fw 190 V 21/U1. There were no plans to rebuild the Fw 190 V 20, which had been badly damaged in a bombing raid on Langenhagen on 5 August 1944, in order for it to take part in the Ta 152 C program.

These plans could not be brought to fruition because of rapid developments in the military situation. Therefore the following reorganization of the test series took place:

1. Fw 190 V 21/U1
   Werknummer 0043, engine test-bed
2. Ta 152 V 6
   Werknummer 110 006, Ta 152 C-0

Front view of the Ta 152 V 7, which was converted to the new integral engine cowling.

[1] Development report Ta 152 C series—standard fighter—dated 24 August 1944, sheet XVII a 1-a 3.

47

3   Ta 152 V 7

Werknummer 110 007, Ta 152 C-0/R11

4   Ta 152 V 8

Werknummer 110 008, Ta 152 C-0 with EZ 42 (lead-computing gunsight system)

These prototypes were to be used for preliminary testing in preparation for the planned Ta 152 C-1 production variant. The prototypes actually earmarked for the Ta 152 C-1 series, the Ta 152 V 10, V 11 and V 12 (WNr. 110 010 – 110 012) were canceled on 18 October 1944.

Two further prototypes for the planned Ta 152 C-3 series, which had a different armament package than the C-1, were:

Ta 152 V 16, We*rknummer* 110 016

Ta 152 V 17, *Werknummer* 110 017.

The Ta 152 V 18, W*erknummer* 110 018, originally planned as the C-4/R11, was canceled on 28 December 1944. Because of conditions in the prototype shop, on 16 January 1945 Focke-Wulf announced that these prototypes were not to be expected before April-May 1945. Therefore, two original Ta 152 H-0 would serve to test the MK 103 engine-mounted cannon planned for the Ta 152 C-3:

Ta 152 V 27, W*erknummer* 150 027

Ta 152 V 28, *Werknummer* 150 030.

It was planned that both would be converted to the DB 603 E with MW 50 instal-

lation,[2] however both machines would retain their original Ta 152 H configuration in order to save time. The following equipment state was anticipated: MK 103 engine-mounted cannon, no weapons in upper fuselage, two MG 151 in wing roots, wing area unchanged at 23.5 m[2].

The Ta 152 V 27, W*erknummer* 150 028, was supposed to be ready to fly on 7 February 1945, the Ta 152 V 28, *Werknummer* 150 030, on 18 February 1945. Chief test pilot Hans Sander flew 150 030 at Langenhagen on 1 and 2 February 1945.

Summary

The DB 603 LA engine planned for the production aircraft was not available for installation in prototypes for the Ta 152 C-1 series at first. Therefore further difficulties were to be expected during installation and testing of the DB 603 LA. The large supercharger air intake caused problems during testing, tearing several times in spite of reinforcement of the bottom of the intake. It was assumed that the cause was a vibration fracture produced by a too narrow cross-section for the pulsating exhaust jet between the bottom of the supercharger air intake and the upper surface of the wing. Nevertheless, during the course of testing there was significantly less criticism of the DB

603 E engine used compared to the Jumo 213 E.

During trials the Ta 152 V 6 (DB 603 E) achieved the following maximum speeds:

Combat power (2,500 rpm, 1.45 ata boost) 547 kph at sea level, 647 kph at maximum boost altitude, emergency power (2,700 rpm, 1.95 ata boost) 617 kph at sea level, 687 kph at maximum boost altitude.

The poor CG position of 0.724 was found to result in unacceptably high instability about the roll axis on the Ta 152 V 6. Not until CG positions of 0.66 to 0.68 could beginning stability be detected. At the same time, stability about the vertical axis became so bad after installation of the ETC 503 that operations by the units with the standard stores rack would have been impossible. A significant improvement could be achieved by using a drop tank with stabilizing fins, however it was doubtful whether this change could have been implemented by the units on a large scale.

Planned Start of Production for the Ta 152 C

The DB 603 E was to be used as a stop-gap at the start of production of the Ta 152 C.

Detail view of CI + XM's fuselage.

Not until after production was under way would the DB 603 LA with its considerably better high altitude performance be introduced. The following factories were earmarked for series production of the Ta 152 C-1:

Ta 152 C-1/R11: ATG Company in Leipzig
Start of production: March 1945
Ta 152 C-1/R11: Siebel Company in Halle
Start of production: March 1945
Ta 152 C-1/R11: MMW-Mimetallwerke Company in Erfurt
Start of production: March 1945

Plans for the GFW and Roland companies to begin production in May 1945 were canceled.

The Ta 152 C-11/R11 was a special variant of the Ta 152 C-1. Since the MMW firm was originally earmarked to produce the Ta 152 E reconnaissance aircraft, which did not come to fruition, it was decided that the E-series fuselages could be used for the C-series. Beginning in April 1945, production was supposed to switch to the original Ta 152 C-1 without camera mounts and fuselage cut-outs. It is known that work had begun in Erfurt on a batch of 30 Ta 152 E-1 reconnaissance machines. Unfortunately there is no information on W*erknummern* of

these machines or the engines used. At least two Ta 152s left the production site in Erfurt-North (see Employment of the Ta 152 H by JG 301). On 15 April 1945 the Americans found two flyable Ta 152s there (including 150 167), one burnt-out Ta 152 and a further 40 Ta 152 fuselages.

There is photographic evidence that production of the C-1 series was also begun by ATG in Leipzig. As well, three Ta 152s without engines were found at the Siebel factory in Halle-Schkeuditz. According to a serial number list dated 7 July 1944, production of the Ta 152 C-1/R11 by Siebel was to be assigned production block 36 (*Werknummern* in the 360 range) and ATG production block 92 (*Werknummern* in the 920 range). Plans to initiate production of the Ta 152 C in March and April 1945 were thrown into disarray by delays in testing the DB 603 LA. Concerning the status of the DB 603 LA, the Fl.E. group of the chief of air armaments issued the following between 22 and 28 January 1945:

Considerable changes required for operational suitability. According to the DB modernization list of 10 January 1945, all engines of the experimental and development series must be modified. Modifications not yet tested in endurance trials.

Blown-up Ta 152 C with DB 603 LA engine at ATG in Leipzig-Mockau after the war.

The Rechlin Testing Station's request for broad-based testing of at least twenty Ta 152 Cs in the manner of the Ta 152 H trials detachment proved impossible on account of the military situation.

Variants of the Ta 152 C

The Ta 152 C was conceived as a medium altitude fighter and fighter-bomber and unlike the Ta 152 H it did not have a pressurized cockpit. The Ta 152 C was also the first Ta 152 which was supposed to be equipped with the new Daimler Benz DB 603 LA engine still under development. Unlike the later planned DB 603 L (9-8603 B1/TL), the DB 603 LA (9-8603 B1/TLA) did not have a supercharger intercooler, therefore methanol-water injection was necessary for the provision of emergency power. This was not the case with the DB 603 L and therefore the MW 50 could be used entirely to boost performance. The DB 603 L developed 1,870 H.P. for takeoff, and this could be increased to 2,250 H.P. As was the case with the Ta 152 H, production of the Ta 152 C was to begin with a pre-production series, the Ta 152 C-0. The pre-production aircraft differed from the production machines in that they lacked the new wing with six bag fuel tanks (470 l). The Ta 152's remaining fuel was carried in the fuselage, with a forward tank holding 233 liters of B4 and a rear tank with a capacity of 362 liters. The 140-liter MW 50 tank was installed behind Bulkhead 8 and was sufficient for about thirty minutes of operation; for thermal reasons, however, use of MW 50 was restricted to three periods of ten minutes. In addition a 300-liter drop tank could be carried. Delays in delivery of the DB 603 LA led to the decision to allow construction of the Ta 152 C-0 to begin with the DB 603 E (power plant 9-8603 B1/TEA). The bad-weather equip-

Specification for the Ta 152 C-1/R14 from the Focke-Wulf Comparison:

Power plant:	DB 603 LA	
	Armed with 2 MG 151/20 (2 x 250 rounds)	
Payload:	LT IB short torpedo 780 kg	
	or LT IB long 850 kg	
	C-1/R14	C-1/R14 with wing tanks
Weight with short torpedo:	5,440 kg	5,780 kg
Weight with long torpedo:	5,500 kg	5,850 kg
Fuel:	592 l B4	1 062 l B4 MW 50:
	140 l	140 l
Return flight at altitude of 3 000 m:	558 kph	607 kph
	552 kph	601 kph
Range:	540 km	1,040 km

Tail of a Ta 152 C, *Werknummer* 500 645, found toward the end of the war.

ment set (*Rüstsatz*) R11, consisting of the FuG 125, the LGW K 23 automatic pilot and heated canopy, was intended for both the Ta 152 C-0 and C-1. Stability problems similar to those encountered by the Ta 152 H led to a reevaluation of the Ta 152 C-1 on 9 March 1945. The tank arrangement was revised, as a result of which the left inner and middle wing tanks with a combined capacity of 150 liters were switched to MW 50 and the 140-liter MW 50 tank located behind Bulkhead 8 was dropped (Ta 152 C-1/ R31). Ta 152 C-1/R11s already built were

to be limited to 115 liters of MW 50 or 280 liters of fuel in the aft 362-liter tank.

All later versions of the Ta 152 C differed only in having modified weapons systems or different radio equipment. While the Ta 152 C-1 was armed with two MG 151/20 in the fuselage, two MG 151/20 in the wing roots and one MK 108 engine-mounted cannon, the Ta 152 C-3 was supposed to be equipped instead with two MG 151/15 in the fuselage, two MG 151/15 in the wing roots and one MK 103 engine-mounted cannon. Instead of the FuG 16 ZY the Ta 152 C-2 and C-4 were to be equipped with the FuG 15. Neither series came to fruition, as it was no longer possible to set up the ground organization necessary for use of the FuG 15.

The Ta 152 C-5's armament initially consisted of five MG 151/20 cannon, but this was later changed. The new weapons set consisted of a MK 103 engine-mounted cannon and one MK 103 in each wing root. This version was later realized as the Ta 152 B-5 powered by the Jumo 213 E. The last version was the Ta 152 C-11/R11, which made use of the fuselage of the E-1 reconnaissance variant under construction by MMW in Erfurt. This version was born of necessity, for the original production sites for the Ta 152 C were unable to begin production because of war developments or had already been largely destroyed.

The Ta 152 C-1/R 14 was the only Ta 152 variant designed as a torpedo carrier by Focke-Wulf. The RLM did not issue the development order for the Ta 152 C torpedo carrier until 12 December 1944. The result was a comparison of the Fw 190 F, Fw 190 D and Ta 152 C-1/R 14, after which Focke-Wulf came out against a version of the Ta 152 C-1/R14 because the reduction in directional stability caused by the large forward lean of the torpedo and the degree of airframe modification were considerably greater than that of the Fw 190 F or Fw 190 D.

Another negative point was the fact that the Ta 152 C had yet to enter series production, which would mean a serious delay in production of the first prototypes. Nevertheless, a trial torpedo installation was made on the Ta 152 V 7 (WNr. 110 007, CI + XM) in March 1945. Focke-Wulf intended to use this mock-up to investigate the aircraft's altered flight characteristics.

The fuel system consisted of two standard fuel tanks in the fuselage with a combined capacity of 592 liters plus a 140-liter MW 50 tank in the fuselage. The aerial torpedo rack (Type 504) was also designed to carry the BT 1400 (Fw 190 C-1/R15). The planned radio equipment: LGW K 23 automatic pilot, FuG 16 ZY, FuG 101a in the wing, and FuG 25a.

Die Varianten der Ta 152 C

Bezeichnung	Motor	Flächen	MW 50	Tank vorn	Tank hinten	Gesamt	Bemerkung
C-0	DB 603 E	–	1401	2331	3621	5951	Vorserie
C-0/R11	DB 603 E	–	1401	2331	3621	5951	Vorserie Schlechtwetterjäger
C-1	DB 603 L/LA	4701	1401	2331	3621	1065l	Vollserie
C-1/R11	DB 603 L/LA	4701	1401	2331	3621	1065l	Regelung bis 9.3.45/ MW 50 - Niederdruckanlage
C-1/R11	DB 603 L/LA	4701	1151	2331	2801	9831	Regelung ab 9.3.45/ MW 50 - Niederdruckanlage
C-1/R14	DB 603 E/LA	4701	1401	2331	3621	1065l	als Torpedoflugzeug geplant
C-1/R15	DB 603 E/LA	4701	1401	2331	3621	1065l	als BT 1400 Trägerflugzeug geplant
C-1/R31	DB 603 L/LA	3201	1501	2331	3621	915l	Regelung ab 9.3.1945'
C-2	DB 603 L/LA	4701	1401	2331	3621	1065l	FuG 15 statt FuG 16ZY, ab 15. Dezember 1944 nicht mehr gefordert
C-2/R10	DB 603 L/LA	4701	1401	2331	3621	1065l	wie Ta 152 C-2, aber mit Aufklärerrumpf Ta 152 E-1
C-2/R11	DB 603 L/LA	4701	1401	2331	3621	1065l	wie Ta 152 C-2, aber mit Schlechtwetterrüstsatz
C-3	DB 603 L/LA	4701	1401	2331	3621	1065l	geänderte Waffenanlage
C-3/R11	DB 603 L/LA	4701	1401	2331	3621	1065l	wie Ta 152 C-3, aber mit Schlechtwetterausrüstung
C-4	DB 603 L/LA	4701	1401	2331	3621	1065l	FuG 15 statt FuG 16ZY, ab 15. Dezember 1944 nicht mehr gefordert
C-4/R11	DB 603 L/LA	4701	1401	2331	3621	1065l	wie Ta 152 C-4, aber mit Schlechtwetterausrüstung
C-5	DB 603 L/LA	4701	1401	2331	3621	1065l	geänderte Waffenanlage FuG 15 statt FuG 16 ZY
C-5/R11	DB 603 L/LA	4701	1401	2331	3621	1065l	wie Ta 152 C-5, aber mit Schlechtwetterausrüstung
C-6	DB 603 L/LA	4701	1401	2331	3621	1065l	FuG 15 statt FuG 16 ZY, ab 15. Dezember 1944 nicht mehr gefordert
C-6/R11	DB 603 L/LA	4701	1401	2331	3621	1065l	wie Ta 152 C-6, aber mit Schlechtwetterausrüstung
C-11/R11	DB 603 L/LA	4701	1401	2331	3621	1065l	Ausführung wie C-1/R11 jedoch Rumpf des Aufklärers »E«

Triebwerksbezeichnungen:

Triebwerk DB 9-8603 B1/TEA-   Motor DB 603 E und MW 50-Anlage

Triebwerk DB 9-8603 B1/TLA   -Motor DB 603 LA-L-Motor ohne Ladeluftkühler mit MW 50-Anlage

Triebwerk DB 9-8603 B1/TL   -Motor DB 603 L mit Ladeluftkühler

Waffenanlagen:

Ta 152 C-1/C-2:   2 x MG 151/20 im Rumpf oben   2 x MG 151/20 in Flügelwurzel   1 x MK 108 als Motorkanone

Ta 152 C-3/C-4: 2 x MG 151/15 im Rumpf oben   2 x MG 151/15 in Flügelwurzel   1 x MK 103 als Motorkanone

Ta 152 C-5/C-6:   2 x MK 103 in Flügelwurzel   1 x MK 103 als Motorkanone

Rüstsatz R11: Der Rüstsatz R11 war eine zusätzliche Ausrüstung für den Schlechtwetterjagdeinsatz und sollte ab 1. Flugzeug sowohl bei der Ta 152 C-0, als auch bei der Ta 152 C-1 zum Einbau gelangen. Er bestand aus dem FuG 125 Hermine, der Jägerkurssteuerung LGW K 23 und Heizscheiben. Bei der Ta 152 C kam keine GM 1-Anlage zur Verwendung.

Änderungen der Baureihenbezeichnungen ab 9.3.1945 wurden erforderlich durch Stabilitätsprobleme.

Focke-Wulf
Flugzeugbau
G. m. b. H.
Bremen

Baubeschreibung Nr. 290

Jagdflugzeug Ta 152 C

Blatt: 7

Focke-Wulf Flugzeugbau G.m.b.H. Nr. 26a

Gewichtsaufstellung

Ta 152

Benennung	C-1 (kg)	C-3 (kg)
Rumpfwerk	384	384
Fahrwerk	245	245
Leitwerk (Metall)	136	136
Steuerwerk	27	27
Tragwerk	557	557
Triebwerk vor Brandschott	1840	1840
Triebwerk in der Zelle	217	217
Normale Ausrüstung	230	230
Zweckausrüstung	365	457
Ballast	13	16
Rüstgewicht	**4014**	**4109**
Flugzeugführer	100	100
Kraftstoff Rumpf vorn	182	182
" Rumpf hinten	283	283
MW 50 140 l im Rumpf hinten	127	127
Kraftstoff im Flügel in 6 Sackbehältern	368	368
Schmierstoff	55	55
Munition 2 MG 151 (2 x 150 Schuß)	66	66
" 2 MG 151 (2 x 175 Schuß)	77	77
" MK 108 (90 Schuß)	50	–
" MK 103 (80 Schuß)	–	75
Zuladung	**1308**	**1333**
Normalfluggewicht	**5322**	**5442**

55

Wichtigste Kenndaten der Bewaffnung

	MG 151/20	MG 151/15	MK 108	MK 103
Kaliber mm	20	15	30	30
Schussfolge Sch/min. ungesteuert	ca.700	ca.700	600	420
Schussfolge Sch/min. bei Steig- und Kampfleistung Luftschr.-Drehzahl = 1300 U/min.	650	650	-	-
Mündungsgeschwindigkeit vo m/sec.				
1.) bei M-Geschoss	805	-	500	900
2.) bei Brandsprenggranate	705	960	-	800
3.) bei Panzergranate	-	850	-	700
Munitionsgewicht gegurtet g/Schuss:				
1.) M-Geschoss	197,2	-	595	920
2.) Brandsprenggranate	223,0	176	-	1200
3.) Panzergranate	223,0	190	-	1055
Reines Waffengewicht kg	42,0	42,0	65,0	145

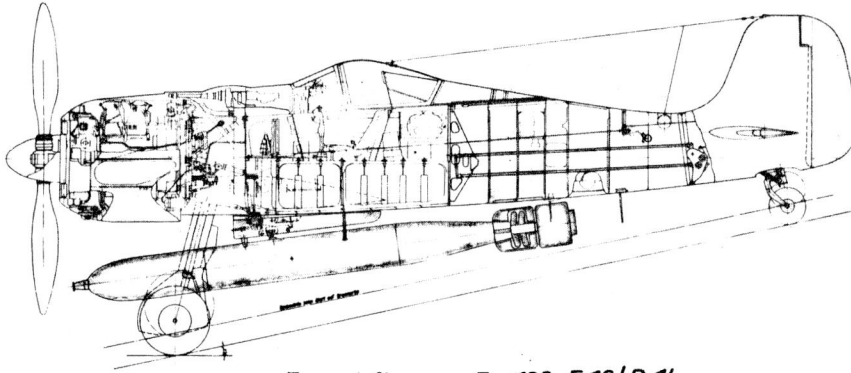

Torpedoflugzeug Fw 190 F 16/R-14
mit LT-Träger (Schloß 504) ohne Torpedosporn

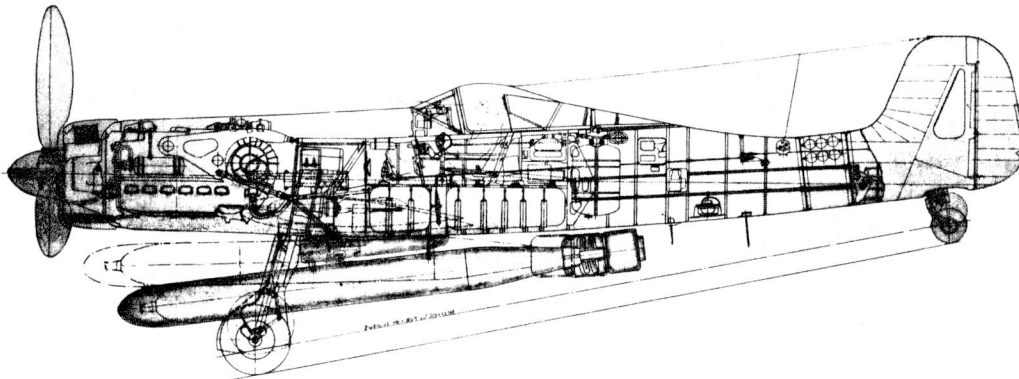

Torpedoflugzeug Ta 152 C-1/R 14
mit LT-Träger (Schloß 504)
mit eingebautem Schloß 504
ohne Torpedosporn

Torpedoflugzeuge Fw 190 F, D, Ta 152 C/R-14

Gegenüberstellung d. LT-Aufhängungen

Focke-Wulf Flugzeugbau G.m.b.H.
24. 1. 44

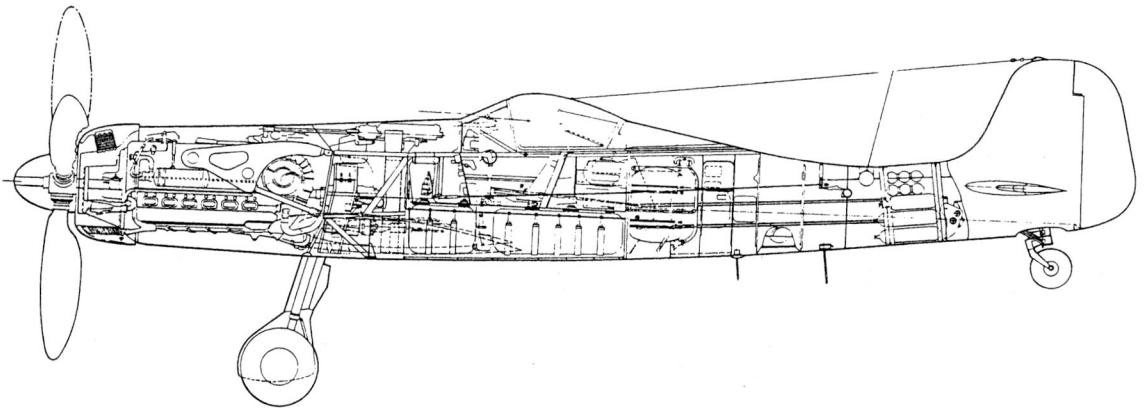

Jagdflugzeug Ta 152 C (Längsschnitt)

mit DB 603 LA Einheitstriebwerk

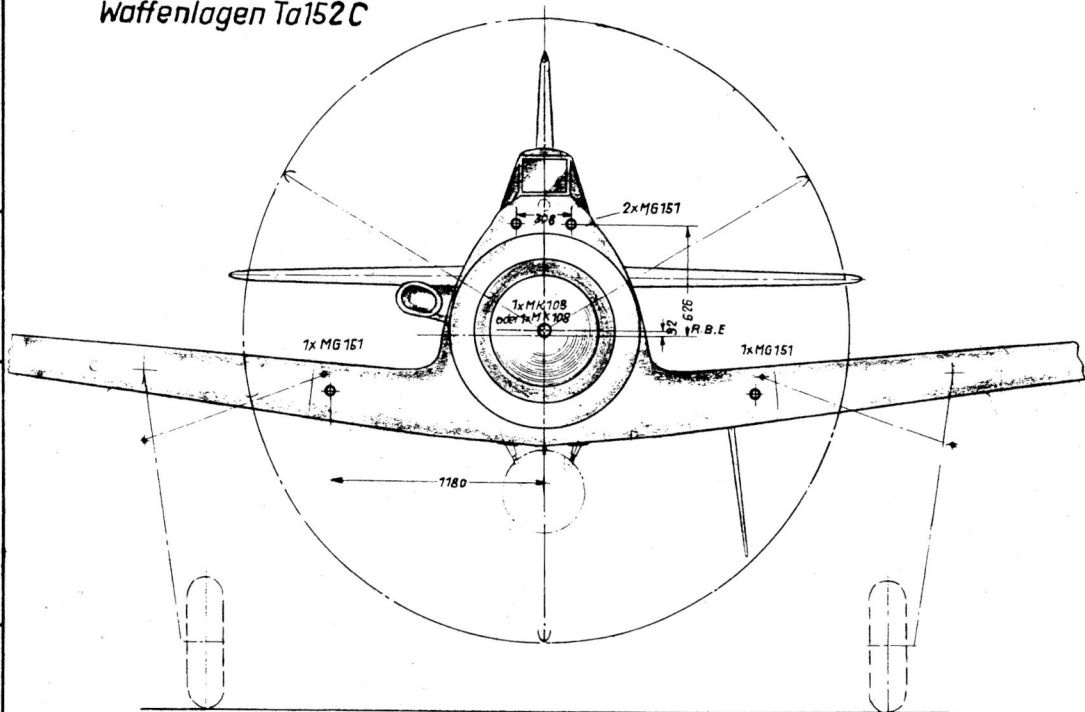

Waffenlagen Ta152C

2x MG151

1x MK108 oder 1x MK108

1x MG151

1x MG151

R.B.E

1180

Focke-Wulf
Flugzeugbau
G. m. b. H.
Bremen

Baubeschreibung Nr. 290
Jagdflugzeug Ta 152 C

Blatt: 9

Mappe Nr.

Ausgegeben

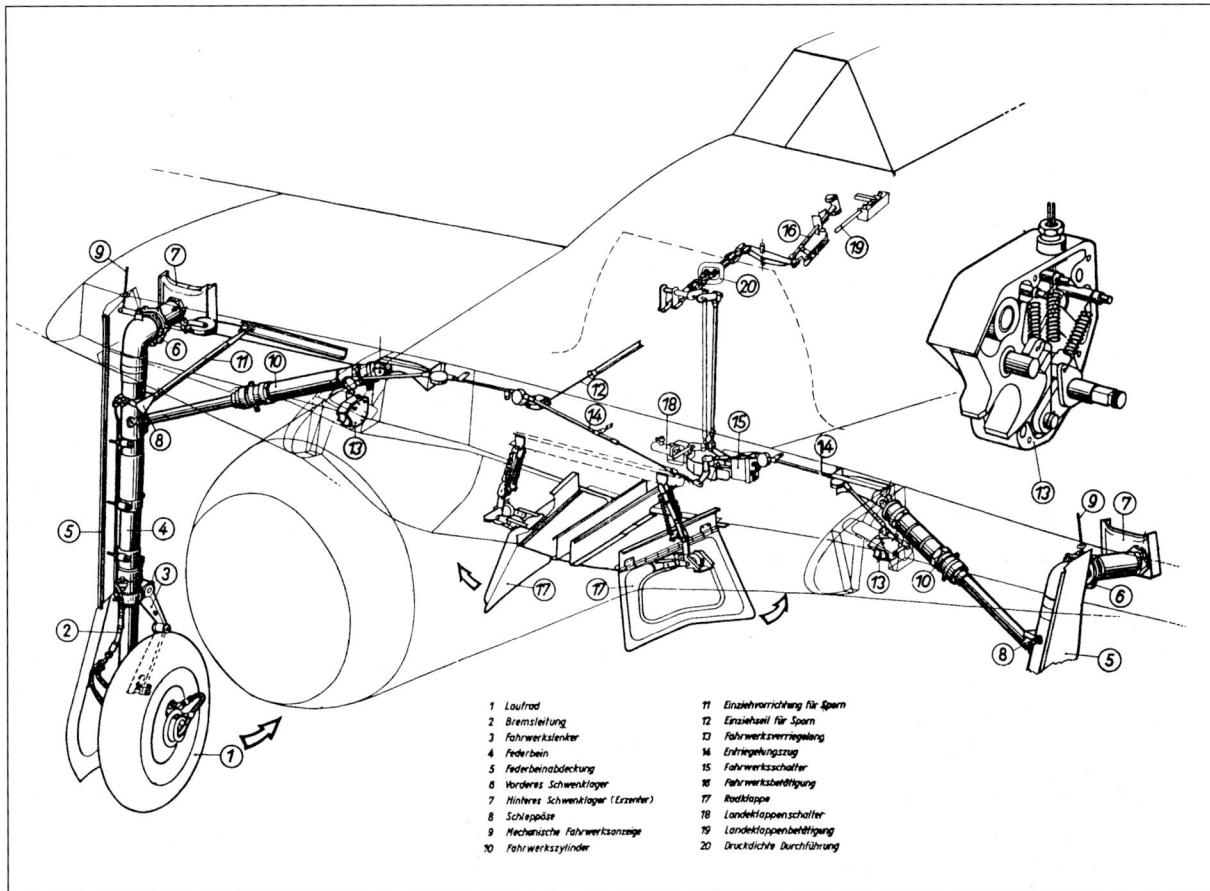

1	Laufrad	11	Einziehvorrichtung für Sporn
2	Bremsleitung	12	Einziehseil für Sporn
3	Fahrwerkslenker	13	Fahrwerksverriegelung
4	Federbein	14	Entriegelungszug
5	Federbeinabdeckung	15	Fahrwerksschalter
6	Vorderes Schwenklager	16	Fahrwerksbetätigung
7	Hinteres Schwenklager (Exzenter)	17	Radklappe
8	Schlepppöße	18	Landeklappenschalter
9	Mechanische Fahrwerksanzeige	19	Landeklappenbetätigung
10	Fahrwerkszylinder	20	Druckdichte Durchführung

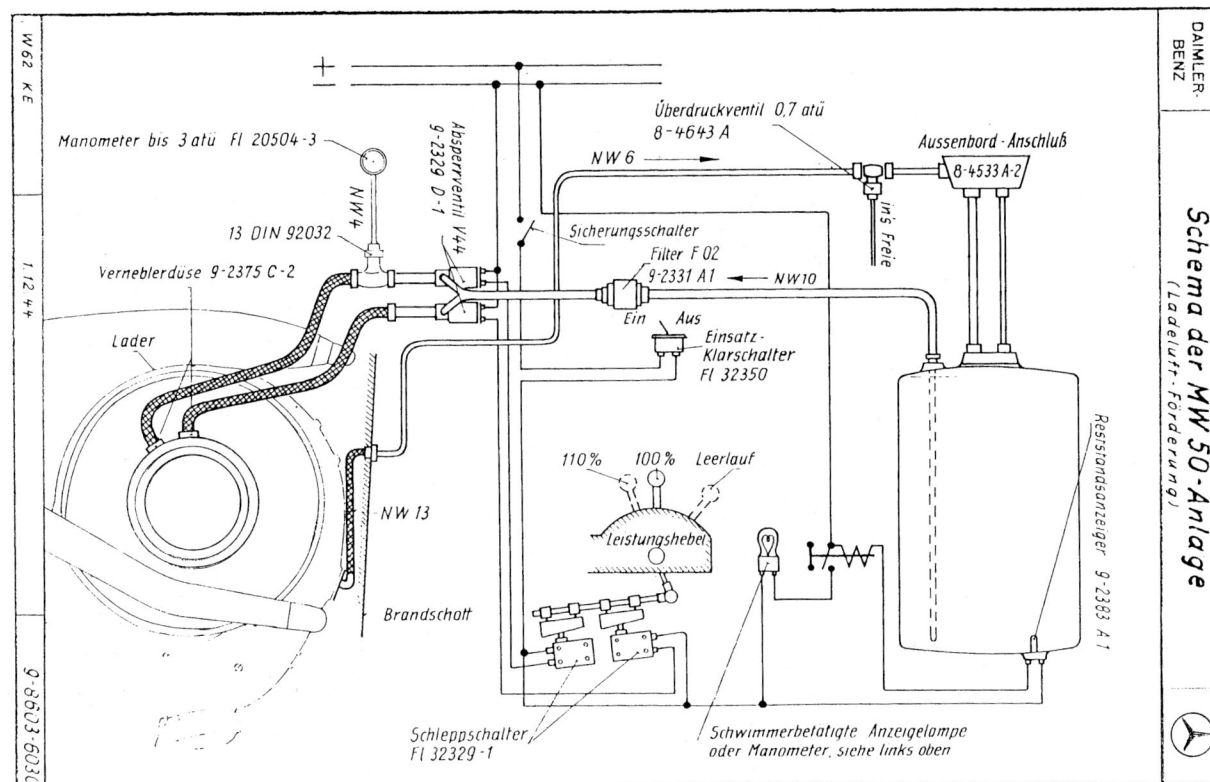

Schema der MW 50-Anlage
(Ladeluft-Förderung)

DAIMLER-BENZ

W 62 K E

1.12.44

9-8603-6030

Manometer bis 3 atü Fl 20504-3

Absperrventil V44 9-2329 D-1

NW 4

13 DIN 92032

Verneblerdüse 9-2375 C-2

Lader

NW 13

Brandschott

Sicherungsschalter

Filter F 02 9-2331 A1 ← NW10

Ein   Aus

Einsatz-Klarschalter Fl 32350

Überdruckventil 0,7 atü 8-4643 A

NW 6 →

ins Freie

Aussenbord·Anschluß

8-4533 A-2

Widerstandsanzeiger 9-2383 A 1

110%   100%   Leerlauf

Leistungshebel

Schleppschalter Fl 32329-1

Schwimmerbetätigte Anzeigelampe oder Manometer, siehe links oben

59

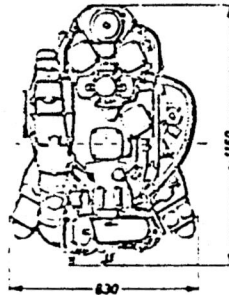

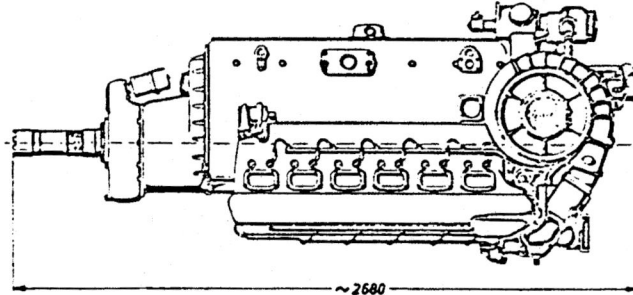

Höhe km	Leistungsstufe	PS	U/min	Ladedruck ata	Kraftstoffverbrauch g/PSh	l/h
0	Start- und Notleistung	1750	2700		235 + 10	565
0	Steig- und Kampfleistung	1580	2500		220 + 10	480
0	Höchstzul. Dauerleistung	1375	2300		215 + 10	410
6,3	Notleistung	1590	2700		235 + 10	520
6,3	Steig- und Kampfleistung	1490	2500		220 + 10	450
6,0	Höchstzul. Dauerleistung	1390	2300		215 + 10	410
5,6	Dauersparleistung	1170	2000		205 + 10	330
10,0	Notleistung	1000	2700			

Untersetzung: E u.F 1:1,93 +)

Vergleichsgewicht: E = 910 kg + 3 %   F = 990 kg + 3 %

Abzuführende Wärmemenge bei Steig- und Kampfleistung:

a. aus Schmierstoff max 55 000 kcal/h am Boden
   "   "   "   40 000 "   in Volldruckhöhe
b. " Kühlstoff   " 420 000 "   "   "

Sonstige Vermerke:
DB 603 E = rechtslaufend
DB 603 F = linkslaufend

Merkmale und Eignung:

Der Motor unterscheidet sich von DB 603 A,D durch:

    größere Volldruckhöhe (Sonderlader)
    höheres Gewicht, längerer Getriebevorbau,
    Bugradbefestigungsaugen, andere Luftschraubenverstellung,
    druckbelüfteter Zündmagnet, angebaute Ölschleuder,
    Benzin-Anlasser, Sturzflugventil,
    Zwischengetriebe für Drehzahlregler, zugleich für Stromerzeugerantrieb,
    Zusatzgetriebe (auf Wunsch)
    angebaute Kühlstoffzusatzpumpe

und ist somit, von Verdichtung und Laderübersetzung abgesehen, ausgeführt
wie DB 603 G,K.

+) Konstruktiv ist auch 1:2,07 möglich, diese aus Beschaffungsgründen
   jedoch nicht freigegeben.

Unterschriften

DAIMLER-BENZ A.G. Stgt.-Untertürkheim 60 AE	R L H GL/C-E3	Mappe Leistungsblätter Ausgabe: 601/LII 4.43
		9-603-6136

60

Steigleistungen über der Flughöhe mit Steig- u. Kampfleistung.

Leistungsvergleich Fw 190 D - Ta 152 C

Horizontalgeschwindigkeit über der Flughöhe mit Steig- u. Kampfleistung.

Leistungsvergleich Fw 190 D - Ta 152 C

Vorteil: Angaben der Leistungsgewinne durch Wasser-Methanol-Zusatz von der Firma Daimler-Benz auch für Steig- u. Kampfleistung als gesichert bezeichnet wird, ergeben sich für die Ta 152 über den gesamten Höhenbereich bessere Flugleistungen gegenüber der Fw 190 D.

Bei der Fw 190 D mit 213 E ist ein Leistungsgewinn durch Wasser-Methanol nicht erreichbar, weil hier der Zusatz als ausschließlicher Ersatz für die Ladeluftkühlung gebraucht wird.

30.5.44

Focke-Wulf Flugzeugbau G.m.b.H.

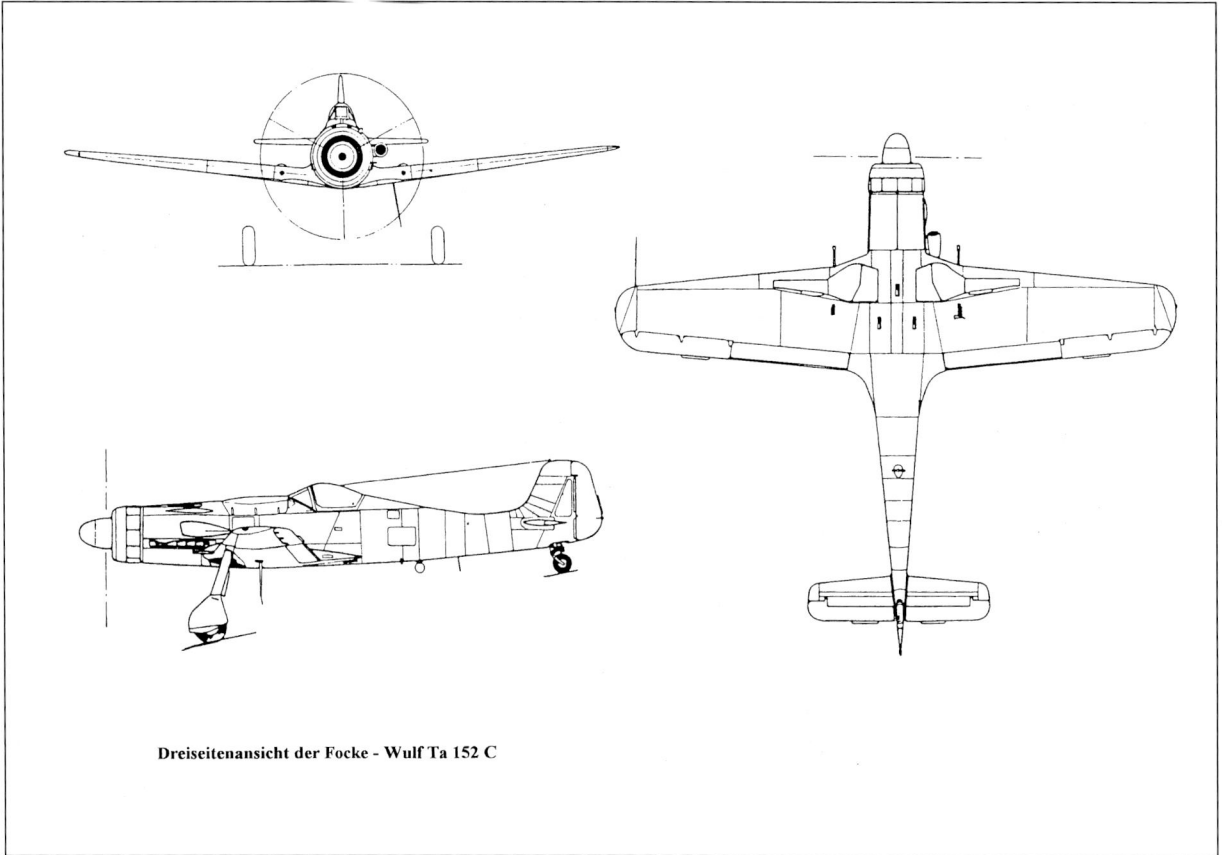

Dreiseitenansicht der Focke - Wulf Ta 152 C

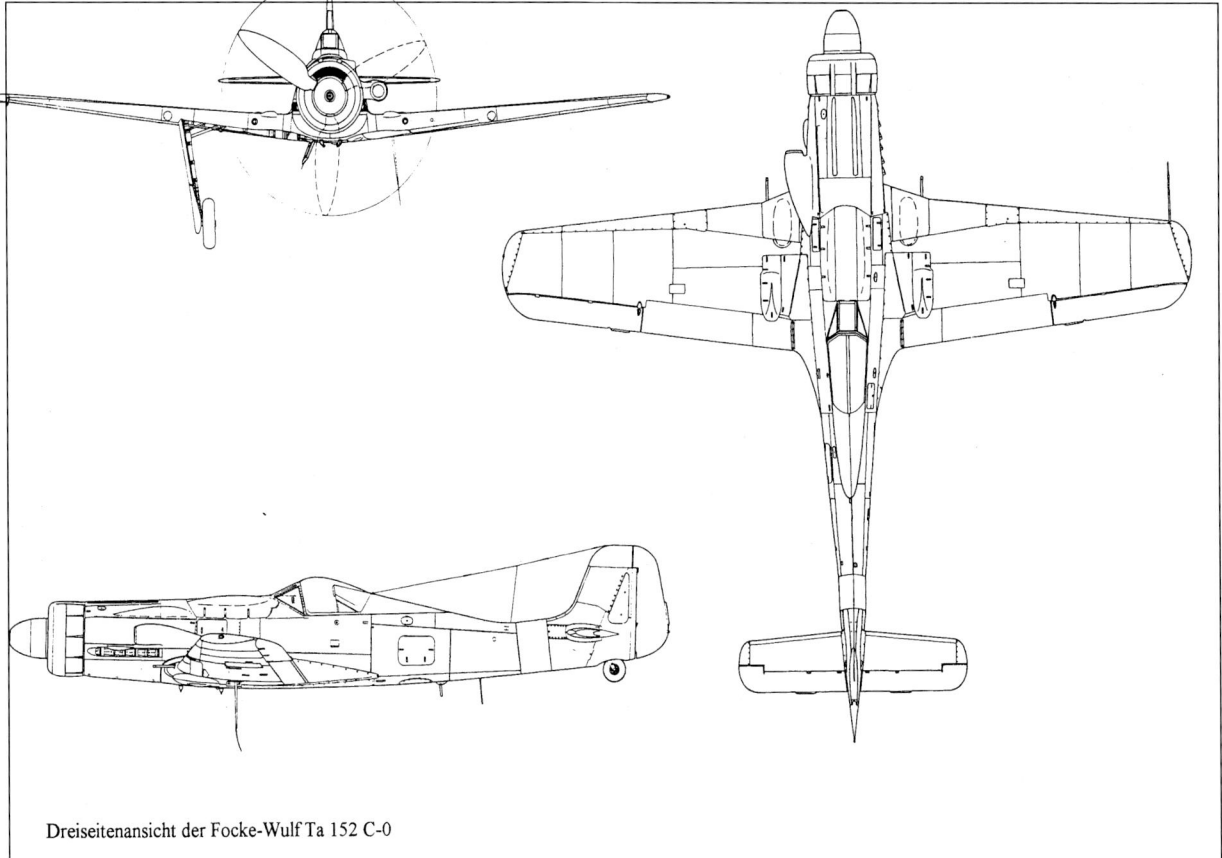

Dreiseitenansicht der Focke-Wulf Ta 152 C-0

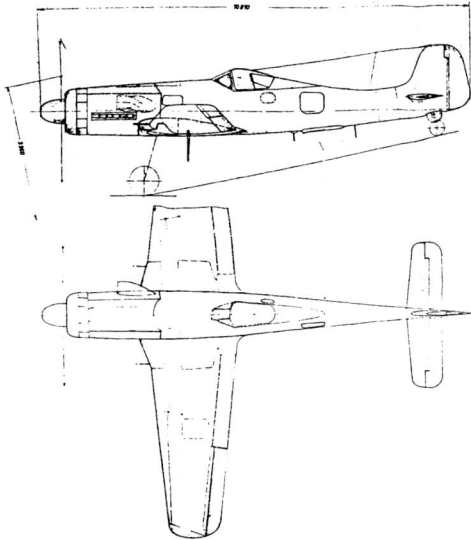

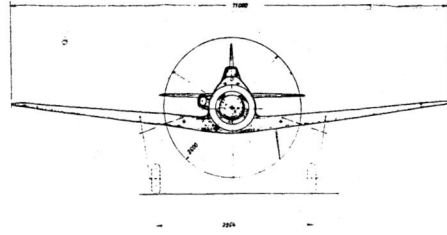

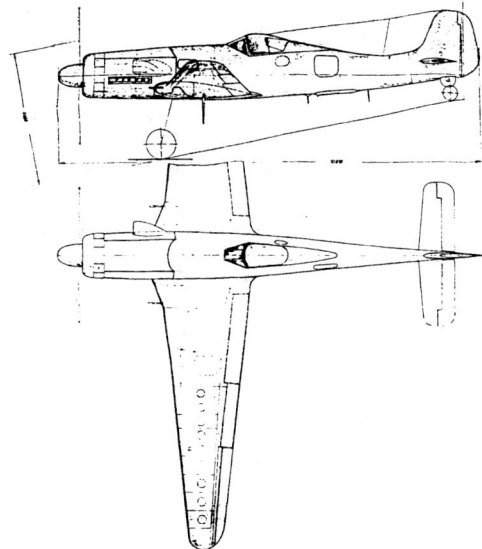

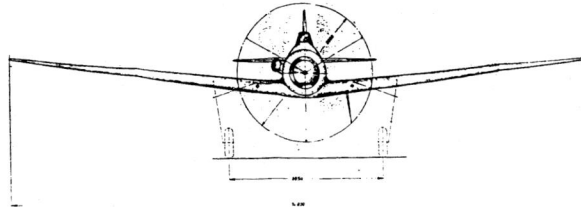

Einheitstriebwerk Jumo 213 E 9

Aufklärer Ta 152 E-1

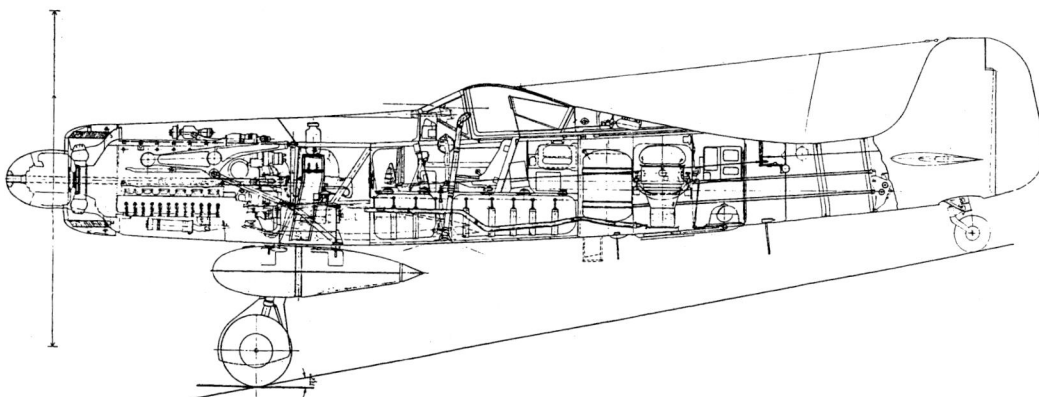

Triebwerk Einheitstriebwerk Jumo 213 E
Bildgerät Rb 75/30   (ohne Flz.V. Anlage)   9

Höhenaufklärer Ta 152 E-2

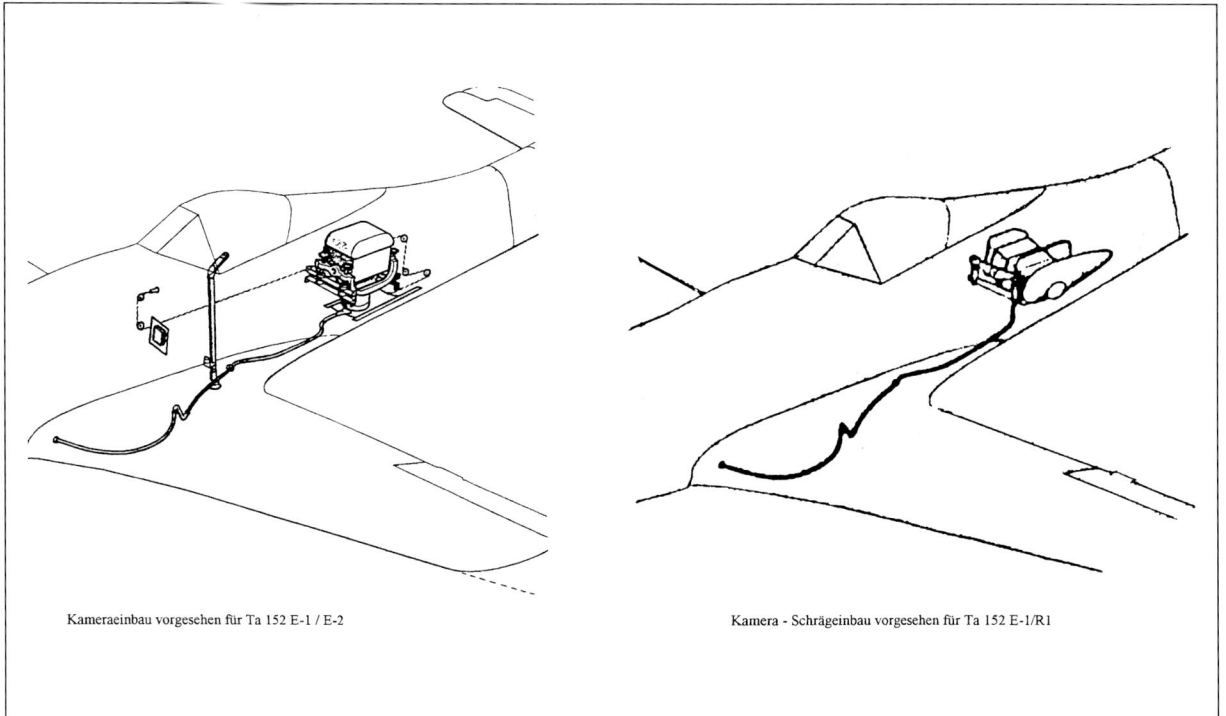

Kameraeinbau vorgesehen für Ta 152 E-1 / E-2

Kamera - Schrägeinbau vorgesehen für Ta 152 E-1/R1

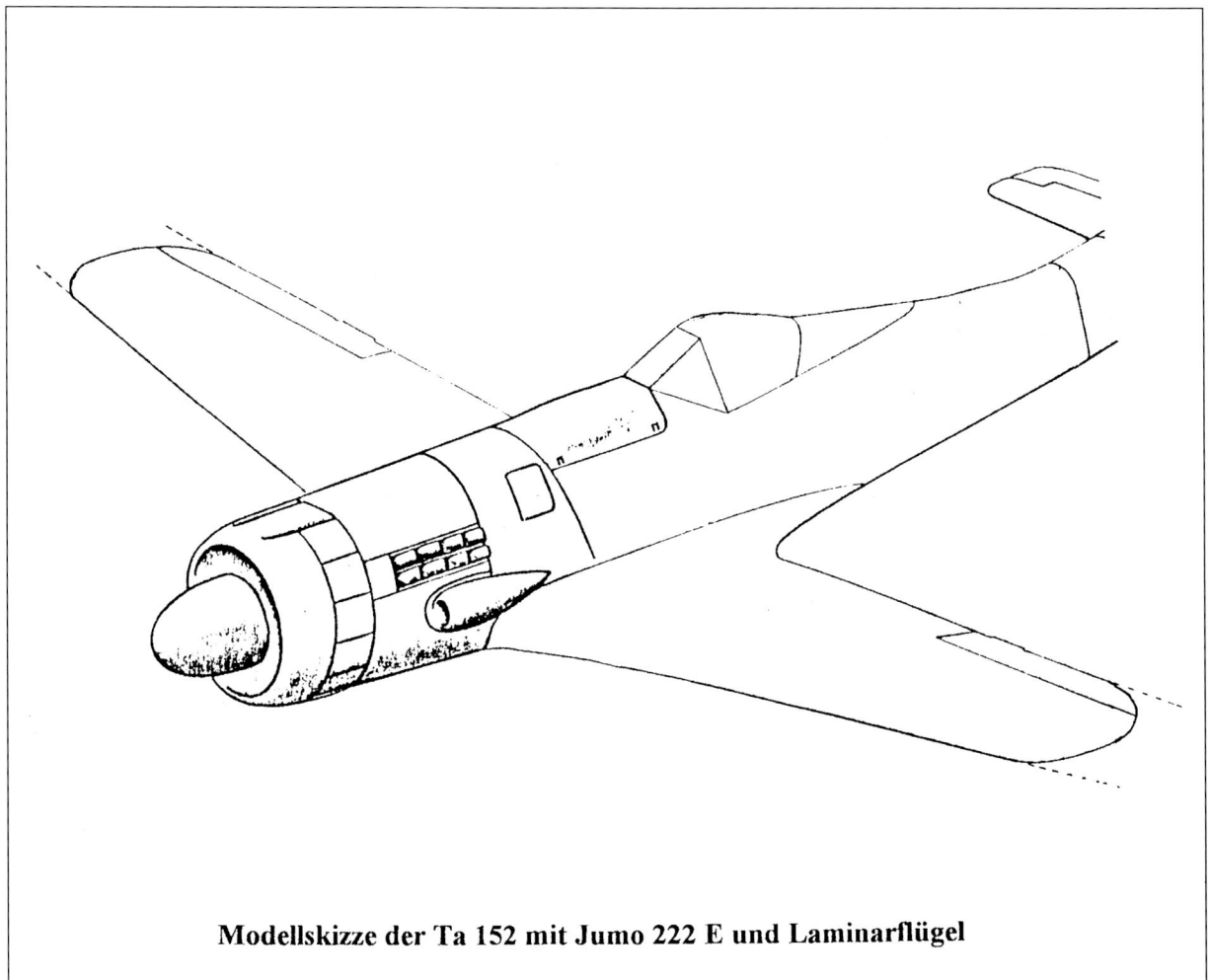

Modellskizze der Ta 152 mit Jumo 222 E und Laminarflügel

C. G e w i c h t s a u f s t e l l u n g

1) Rüstgewicht

Rumpfwerk	404 kg
Fahrwerk	246 kg
Leitwerk	125 kg
Steuerwerk	36 kg
Tragwerk	818 kg
Triebwerk	2476 kg
Zweckausrüstung {2 MG 151 i.Rumpf {2 MG 151 i.Flügel	261 kg
Normalausrüstung	237 kg
Rüstsatz (MW 50 Behälter im Rumpf oder Kraftstoffzusatzbehälter)	15 kg
Rüstgewicht	4618 kg

2) Zuladung

Führer	100 kg	100 kg
Kraftstoff im Rumpf	440 kg	440 kg
Kraftstoff im Flügel	- kg	354 kg
Schmierstoff	40 kg	61 kg
MW 50 - Füllung	- kg	125 kg
Munition (Rumpf 2 x 150 Schuß)	54 kg	54 kg
" (Flügel 2 x 175 Schuß)	63 kg	63 kg
Kraftstoff im Rumpfzusatzbehälter	85 kg	- kg
Zuladung	782 kg	1197 kg

3) Zusammenstellung

Rüstgewicht	4618 kg	4618 kg
Zuladung	781 kg	1197 kg
Bemessungsgewicht	5400 kg	- kg
Abfluggewicht	- kg	5815 kg

Focke-Wulf Flugzeugbau G.m.b.H. Nr. 26 a

Development of the High-altitude Fighter Focke-Wulf Ta 152 H

Focke-Wulf's experience in building high-altitude fighters ("*Höhenjäger 1*" and "*Höhenjäger 2*") and the RLM's demands for a suitable high-altitude fighter led to its development of the Focke-Wulf Ta 152 H high-altitude escort fighter.

The opportunity for Focke-Wulf to develop the Ta 152 H resulted from the failure of the Me 109 H high-altitude fighter, which was essentially a modified Bf 109 G. The RLM demanded a service ceiling of 13,000 to 15,000 meters. The aircraft's primary role was to be the interception of high-flying enemy reconnaissance machines. Testing began with the Bf 109 V 54 (PV + JB) on 5 November 1943. The insertion of a rectangular center section resulted in an increased wingspan of 13.26 meters (wing area 21.90 m²). The horizontal tail was also increased in area and the track of the critical Bf 109 undercarriage was widened. During the course of trials, in which the Bf 109 V 55 (PV + JC) also took part, the Bf 109 H achieved a ceiling of 14,300 meters and a maximum speed of 580 kph at 12,600 meters with GM 1 injection. The Bf 109 V 54 displayed poor flight characteristics about all three axes during flight tests. In addition, the enlarged wing tended to flutter as a result of unforeseen vibration. The Bf 109 H was canceled by the RLM on 18 July 1944, along with plans to construct the Bf 109 H-2, based on the Bf 109 K-4/R2. The Me 262 had absolute priority by this time and the new Bf 109 K-4 offered sufficient high-altitude flight characteristics. On 7 December 1943 the RLM instructed Focke-Wulf to construct six prototypes for the planned Ta 152 H high-altitude fighter. As in the case of the Bf 109 H, the RLM demanded that the Ta 152 H be produced from the series airframe of the Fw 190 A-8 with a minimum of changes, so as to enable the use of as many existing jigs as possible. The Fw 190 V 33/U1 (GH + KW) made its first flight on 13 July 1944.

The Ta 152 H was essentially a high-altitude version of the Ta 152 C fighter aircraft, differing from the latter mainly in having a larger wing with improved aspect ratio and a pressurized cockpit. The special demands of high-altitude operations were to be met by the installation of both GM 1 and MW 50 systems as standard equipment. Armament was limited to a MK 108 engine-mounted cannon and two MG 151/20 cannon in the wing roots. The anticipated power plant was the Junkers Jumo 213 E high-altitude engine. In this case, too, the Daimler Benz DB 603 LA and DB 603 L were planned as alternate power plants from the beginning with the minimum possible engine and airframe changes.

Fuselage

Essentially the following changes were made to the fuselage compared to that of the Fw 190 A-8:

A lengthening of the forward fuselage by 772 mm was necessary as a result of the greater space required by the engine-mounted MK 108. The fuselage extension was the same length as that of the Ta 152 C for reasons of airframe standardization. This allowed the MK 103 to be installed in place of the MK 108. In order to minimize the procurement of new jigs to an absolute minimum, the fuselage extension was bolted directly to the existing engine attachment points. The wing, which was moved forward 420 mm for center of gravity reasons, was attached in the center of the extra

Technical Specification No. 292 Ta 152 H from 15 January 1945

Data Sheet for the Ta 152 H-0 and H-1

Purpose:	Single-seat fighter with pressurized cockpit
Configuration:	Single-engined, low-wing cantilever monoplane with hydraulically retractable undercarriage
Structural strength:	Load coefficients 5.0 and –2.5 at a design weight of 4,500 kg
Power plant:	Jumo 213 E with supercharger intercooler, performance increase by MW 50 and GM 1 systems.

Dimensions:

Wing area	23.3 m²	
Wingspan	14.44 m	
Aspect ratio	8.93	
Vertical tail area	1.77 m²	
Horizontal tail area	2.28 m²	
Maximum length	10.71 m	
Maximum height	3.36 m	
Normal takeoff weight:	Ta 152 H-0	4 730 kg
	Ta 152 H-1	5,220 kg

Armament:
Ta 152 H-0 and H-1
2 MG 151/20 in wing roots with 175 rounds per gun
1 MK 108 engine-mounted cannon with 90 rounds

Armor:			
	Engine armor	10/5 mm	62 kg
	Cockpit armor	20, 15, 10, 8, 5 mm	88 kg
	Armored windscreen 70 mm		
	Total weight of armor		150 kg

Equipment:
FuG 16 ZY, FuG 25a, FuG 125, K 23 autopilot, Revi 16b
300/10 cockpit air compressor with regulating equipment
EZ 42 to soon replace Revi 16b

Fuel system:
Ta 152 H-0 (start of production)
Normal in fuselage 594 l B4, 115 l B4 in extra tank
Drop tank under fuselage 300 l B4
Ta 152 H-1
Normal in fuselage 594 l B4, 85 l GM 1
Additional 5 wing tanks 400 l B4, 1 wing tank with 70 l MW 50
Drop tank under fuselage 300 l B4

section. At the same time, repositioning the wing made it necessary to relocate the rear spar junction and the corresponding fuselage bulkhead. The resulting revised location of the forward fuel tank made it necessary to redesign the fuel tank compartment cover and fuselage sides in the affected area.

In order to avoid having to accept a reduction in stability, especially directional stability as a result of the lengthened engine compartment, the aft fuselage was lengthened by inserting a 0.5-meter-long cylindrical section. The latter also served to accommodate the oxygen bottles and compressed air bottles for the engine-mounted cannon which had been moved aft for center of gravity reasons. The increased fuselage

moment resulting from the lengthened fuselage made it necessary to strengthen the frame assembly. This strengthening was accomplished by fitting steel extrusions instead of the Dural extrusions previously used. This strengthening was identical in scope to that of the Ta 152 C.

Unlike the Fw 190 A and Ta 152 C, the fuselage center section was designed as a pressurized cockpit. The pressure chamber, which had a volume of about one cubic meter, comprised the area above the tank installation. Sealing of the skinning was accomplished using DHK 8800 paste, which was applied to the rivet surfaces. A closer rivet pattern was also planned. The sliding hood was sealed by means of a circular tube. Partly filled with foam rubber, it was

from the tube, then the lock was released and the jettisoning procedure initiated. For de-misting reasons, the windscreen was designed as a double-pane assembly with the following Plexiglas thicknesses: outer 8 mm, inner 3 mm. The inner and outer panels were separated by a space of 6 mm. The air between the panels was dried by eight Silicagel capsules. Lead-ins through the pressurized walls were achieved as follows:

1 Electrical lead-ins: AEG conduit.
2 Hydraulic lead-ins: double flange tubes.
3 Elevator control: rotating conduit with radial seal.
4 Rudder control rods: box gland.
5 Aileron control rods: rubber cub.
6 Engine control rods: pushrods with Junkers conduit.

The engine compartment hatch was sealed by means of a foam rubber ring and was activated by a central latch. The weapons access hatch at Bulkhead 1 was of a similar design. This hatch was also installed on the Ta 152 C, as in both cases it also served as a gas seal. As designed, the Ta 152 C fuselage was largely ready to accept a

Left side panel in the cockpit of the Ta 152 H. Clearly visible is the GM 1 switch T 7 (39) above the throttle lever (36). pumped up to 2.5 atmospheres by a 1-liter compressed air bottle by way of a pressure reducer valve. If jettisoning of the hood became necessary, the air was first evacuated

Cockpit of the Ta 152 H with turn and bank indicator/artificial horizon (7), rate of climb indicator (8) and compass (9). Beneath the compass are the switches for lowering the undercarriage (19) and landing flaps (20) and, from the left, the oil pressure gauge (24) and fuel (25) and oil pressure indicators (26).

pressurized cockpit, consequently the only differences in the Ta 152 H fuselage were the sealing measures described above. The associated oxygen supply is described in detail under the heading "Equipment."

Undercarriage

The undercarriage leg including shock strut and mounting was adopted from current Fw 190 A-8 production. The former electric drive was changed to a hydraulic system. Larger 740 x 210 wheels were installed on account of the aircraft's increased takeoff weight.

Tail Surfaces

Only the horizontal stabilizer and elevator were retained from the Fw 190 A series. The fin and rudder were enlarged for reasons of directional stability. The reinforced standard tailwheel with tire measurements of 389 x 150 mm was installed in conjunction with the larger fin and rudder. In order to achieve adequate longitudinal stability, the vertical and horizontal tails, which together with the rear fuselage formed a single component, were moved aft through the insertion of a 0.5-meter-long cylindrical section at the attachment point. Soon, however, construction of the aft fuselage was switched from Dural to wood. The shape of the tail surfaces remained exactly the same. The fuselage insert was dropped. Instead the extension was incorporated into the aft fuselage, resulting in an improved shape. The extended wing resulted in new ailerons and landing flaps. The flaps were changed from electric to hydraulic operation.

Areas of Aerodynamic Surface
Vertical tail: 1.77 m²
Horizontal tail: 2.82 m²
Ailerons: 2 x 0.56 m²
Landing flaps: 2 x 1.36 m²

Flight Control System

The control system remained essentially unchanged, however the more forward engine position and the fuselage extensions resulted in some changes to the linkages. The pressurized cockpit made it necessary

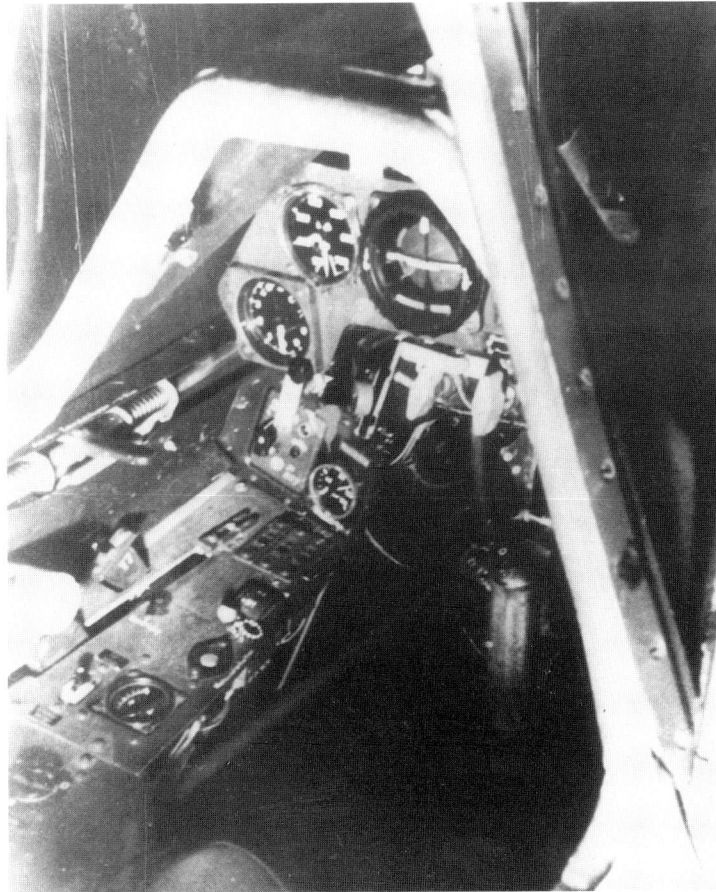

to adopt pressure-tight conduits for the control linkages (see Fuselage). Other changes resulted from the installation of wing tanks.

Wing Assembly

In order to provide propeller clearance from the larger wheels, it was necessary to move the latter outboard by 250 mm. While the inner wing structure was retained largely intact, a 0.5-meter-long spar section was inserted in the center of the wing. This made it necessary to redesign the wing-fuselage junction to the same extent as on the Ta 152 C. In spite of a reduced load coefficient of 5.0, the increased wingspan (14.4 m) and wing area (23.3 m²) made it necessary to increase wing skin thickness. The basic design of the wing was retained, namely a monocoque structure with a forward spar designed as a transverse force bearer. The transverse force was directed over the rear spar and the wing leading edge, which was reinforced by a stiffening rib between each pair of full ribs. The departure from the

Ta 152 H instrument panel. Above and to the right of the altimeter (5) is the airspeed indicator (6).

Detail view of the new, larger supercharger air intake for the Jumo 213 E of the Ta 152 H, seen here on the Fw 190 V 30/U1.

chord-separated mode of construction made necessary the installation of numerous access holes on the wing underside. These were used for assembly and repairs.

The installation of three bag tanks made necessary modification and partial strengthening of the wing in the affected area. Access holes with diameters of 200 mm were installed in the skinning on the bottom of the wing for installation of the tanks. In order to facilitate wing repairs, a separation point was installed in the previ-

Power Plant

The Jumo 213 E high altitude engine developed by Jumo was installed in the Ta 152 H.

Power plant:	Liquid-cooled, 12-cylinder in-line engine with two-stage supercharger
Reduction:	1 : 2.40
Take off power:	1,730 H.P. (1272 kW) at 3,250 rpm at sea level
Climb and combat power:	1,580 H.P. (1162 kW) at 3,000 rpm at sea level
Climb and combat power:	1,260 H.P. (927 kW) at 3,000 rpm at 10 700 meters
Fuel:	87 octane B4, further development with C3 (100 octane) planned
Radiator:	Inner radiator with radial flow-through consisting of four segments. Radiator frontal area 65 dm^2
Lubricant cooling:	Heat exchanger
Cooling system:	Coolant radiator and heat exchanger lie one behind the other in the main cooling circuit. The supercharger air cooler is so arranged in the secondary circuit that coolant is taken from the water pump discharge and is fed to the coolant pump intake via the supercharger air heat exchanger.
Exhaust system:	Normal ejector exhausts
Propeller:	3-blade Junkers variable-pitch propeller with VS 9 wooden blades D = 3.6 m, 1/D = 12.2%. The VS 9 will later be replaced by the 4-blade VS 19 capable of greater power input. D = 3.5 m, 1/D = 11.5%

ously one-piece structure. The separation point consisted of wedge-shaped butt straps which were bolted to the top and bottom flanges of the forward spar.

Alternate Power Plants

The DB 603 LA or DB 603 L were seen as back-up solutions to the Jumo 213 E. A change of power plants would require several minor changes to the airframe and engine accessories (control rods, cockpit air compressor, propeller pitch control, etc.).

Power Boosting

A methanol-water system (MW 50) was planned for increased performance below the maximum boost altitude. The fluid was held in the left inner wing tank, which had a capacity of 70 liters. Use of MW 50 produced an increase in boost pressure. MW 50 injection also served the purpose of internal cooling, to avoid damaging thermal stresses. With a rate of consumption of 150 l/hr

the supply of 70 liters provided an operating duration of about 28 minutes. A GM 1 system was planned for increased performance above the maximum boost altitude. 85 liters of fluid was contained in a circular tank in the aft fuselage. The increase in engine output above the maximum boost altitude was up to 410 H.P. (302 kW). With an average consumption of 100 g/sec, the operating duration of the GM 1 system was a good 17 minutes. Rates of consumption: 60, 100, 150 g/sec.

Fuel System

It was possible to retain the 233-liter forward fuel tank from the Fw 190 A series unchanged, however it was moved forward in keeping with the revised wing position. As a result it was possible to increase the capacity of the rear tank by 70 liters to a total of 362 liters. This increased standard tankage to a total of 595 liters. The tanks were protected tanks with following wall thick-

71

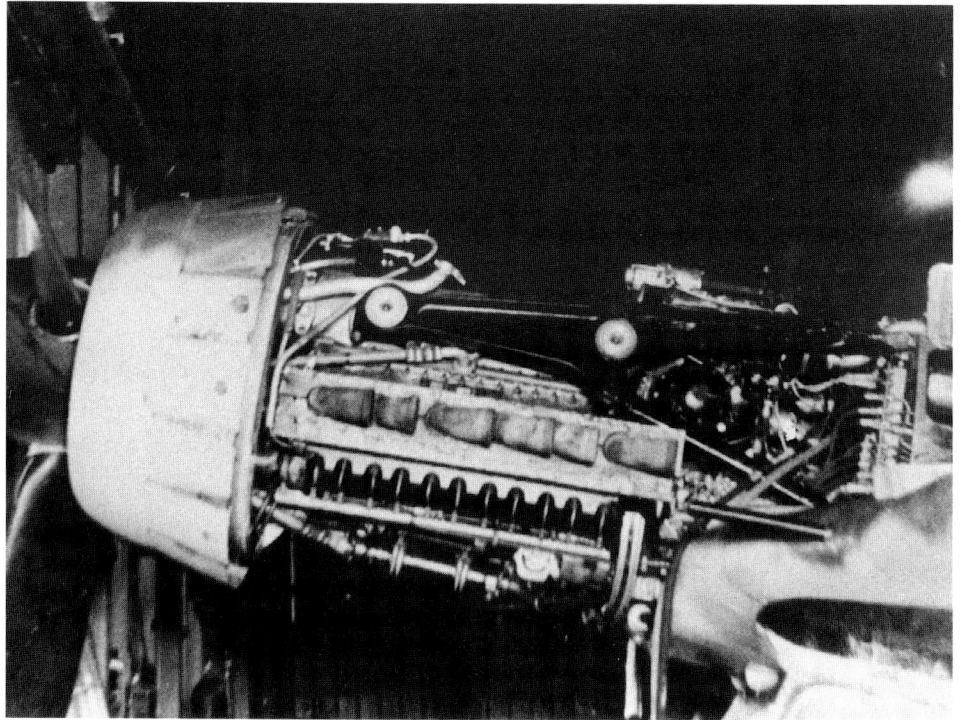

Jumo 213 E engine installed in the Fw 190 V 33/U1, *Werknummer* **0 058, GH + KW, first prototype of the Ta 152 H.**

nesses: sides and bottom 16 mm, top 12 mm. For increased range the following auxiliary tank installations were planned:

Ta 152 H-0

A 300-liter drop tank which was attached to an aerodynamically faired external mount, the so-called "Ta 152 Tank Carriage," by grommets. Fuel transfer by supercharger air.

Ta 152 H-1
(see Ta 152 Weapons and Fuel System)

Six unprotected bag tanks in the wing with a total capacity of 470 l. The left inner bag tank with a capacity of 70 liters was used as the MW 50 tank. Fuel transfer by supercharger air. The tanks were installed through access ports in the underside of the wing (see Wing).

For longer-range missions the 300-liter drop tank (eventually a 600-liter tank) described above could be used.

Lubrication System
The oil tank, with a total capacity of 72 liters, was installed on the right side of the forward fuselage extension next to the engine cannon. It was an unprotected aluminum tank and was largely protected against fire from ahead by the engine. The tank's capacity of 61 liters was sufficient for operation with 594 liters of fuel in the fuselage tanks and a 300-liter drop tank with 25% cold start mixture.

Standard Equipment
Equipment was taken from the Fw 190 A-8 series. Changes included the hydraulic systems made necessary by the change to hydraulic undercarriage and landing flaps as well as minor modifications resulting from the installation of the Jumo 213 A.

Other changes were caused by the cockpit pressurization system described below.

Cockpit Pressurization

A Knorr 300/10 air compressor was installed to provide the pressurized cockpit with aspirable air. It was bolted to the engine with no intermediate gearing. The system functioned as follows:

The aspirable air compressor drew air from the air scoop in front of the radiator and forced it through a filter, a non-return check valve and a modulating piston into the cockpit. When the compressor was switched off, the non-return check valve closed off the line to the compressor and prevented the cockpit air from streaming out through the compressor. The pressurization system began operating at an altitude of 8,000 meters. Above this height cockpit pressure was maintained at a constant 0.36 atmospheres by means of a back pressure regulating valve. At a pressure of 0.23 atmospheres the over-pressure safety valve kicked in, protecting the pressure chamber from excess static pressure. At altitudes below 8,000 meters, with the cockpit unpressurized, fresh air was vented through the air scoop via a non-return check valve, which closed if the cockpit was pressurized. A slide valve allowed fresh or pressurized air to be directed into the cockpit as required, enabling temperature to be regulated to a certain degree.

It was anticipated that heating of the cockpit during pressurization would probably be held within bearable limits. If flight tests revealed unbearable heating at high altitudes, plans would have to be made for back cooling the compressed air.

Ta 152 Armor

Type	Armor Thickness (mm)	Armor Weight (kg)
1) Forward circular armor (engine)	15	39.0
2) Rear circular armor (engine)	8	22.5
3) Armor in front of windscreen	15	14.0
4) Armored windscreen	70	22.5
5) Back armor	8	18.2
6) Shoulder armor	5	5.9
7) Armor plate on Bulkhead 5	5	7.9
8) Head armor	20	20.0
Total weight of armor		150.0

Further strengthening of the armor was planned, especially the back armor to a thickness of 15 mm.

Left side view of the Fw 190 V 33. Clearly visible are the Jumo 213 E, engine mount and supercharger (Bau 8).

The most significant planned systems were:

- FuG 16 ZY radio (transmit and receive)
- FuG 25a IFF set
- FuG 125 navigation equipment, beam-riding method
- Remote reading compass
- Turn and bank indicator and the usual monitoring and navigation equipment
- LGW K23 automatic pilot
- Knorr 300/10 aspirable air compressor
- Revi 16b, is being replaced by the EZ 42 aiming system with automatic lead computing

Specific Equipment

Ta 152 H-0 and H-1
2 MG 151 (20-mm caliber) in wing roots with 175 rounds per gun
1 MK 108 engine-mounted cannon with 90 rounds

Of the above weapons the two MG 151 fired through the propeller disc and were controlled electrically. The carriage of various types of rocket weapon beneath the wings was possible. For ballistic reasons the MK 108 and the entire power plant were aligned approximately 35 minutes positive.

Gravity Weapons

There was no provision for carriage of bombs under the wings or fuselage since the Ta 152 was developed as a pure escort aircraft. The Ta 152 Tank Carriage described under Fuel System was used when a 300-liter drop tank was installed.

Passive Protection

Cockpit armor was expanded and strengthened in keeping with the increased demands resulting from the heavier armament carried by Allied machines.

Structural Strength

The Ta 152 strength manual (memo dated 25 May 1943) prescribed a wing load coefficient of +6.5 or −2.5 for the design takeoff weight of 4,400 kg. The components shared with the Ta 152 C, such as the fuselage and empennage, had considerable strength reserves since these were designed for the demands of the standard fighter role. The engine mount was similar to that of the Ta 152 E (reconnaissance aircraft) and had a load coefficient of 6.5.

The Prototypes Fw 190 V 18/U2, V 29/U1, V 30/U1, V 32/U1, V 33/U1 And Ta 152 V 25

Focke-Wulf revealed the planned series and prototypes in its Ta 152 series overview of December 1943. For the Ta 152 H-1, three prototypes in the original Ta 152 H-1 configuration were planned: Ta 152 V 3 (WNr. 260 001), Ta 152 V 4 (WNr. 260 002) and Ta 152 V 5 (WNr. 260 003). But at that time it was already clear that completion of these prototypes could not be expected before August 1944. Focke-Wulf therefore decided to convert the Fw 190 V 33 at Adelheide. As a result of further delays, Focke-Wulf decided to abandon construction of the three original Ta 152 H-1s. Instead a total of six prototypes were to be built for preliminary testing, with five of them being conversions of existing test-beds. The purpose was to carry out power plant trials with the new Junkers Jumo 213 E engine and investigate the aircraft's handling qualities as quickly as possible.

After the decision by Focke-Wulf on 23 August 1944 the following prototypes were planned and built:

View from the assembly hall at Adelheide. A Fw 190 D-11 (V 55 or V 56) is rolled from the assembly hall; the Fw 190 V 30/U1, GH + KT, is parked on the airfield.

Another photo taken during the roll-out of the Fw 190 D-11 with the registration GH + KT. In the foreground the conversion of further new prototypes goes on (summer 1944).

Fw 190 V 33/U1
Werknummer: 0 058
Registration: GH + KW
First flight: 13 July 1944
Fate: 13 July 1944: 70% write-off in crash at Vechta during ferry flight from Adelheide to Langenhagen

Fw 190 V 30/U1
Werknummer: 0 055
Registration: GH + KT
First flight: 6 August 1944
Fate: 23 August 1944: crashed while on approach to land, 100% write-off. Focke-Wulf pilot Alfred Thomas died in the crash.

Fw 190 V 29/U1
Werknummer: 0 054
Registration: GH + KS
First flight: 24 September 1944
Fate: unknown

Fw 190 V 18/U2
Werknummer: 0 040
Registration: CF + OY
First flight: 19 November 1944
Fate: Parked at Reinsehlen air base on 6 April 1945 and later blown up.

Fw 190 V 32/U1
Werknummer 0 057
Registration: GH + KV
First flight: not before 30 January 1945
Fate: Parked at Reinsehlen air base

All five of these prototypes originated from the "*Höhenjäger 2*" test program. The "*Höhenjäger 2*" was Focke-Wulf's attempt to increase the speed of the Fw 190 at high altitude by the first use of an exhaust-driven turbo-supercharger. The Fw V 29, V 30, V 31, V 32 and V 33 had been specially converted for this program with the DB 603 S (A) engine and the exhaust-driven turbo-supercharger, while the Fw 190 V 18/U1 took part in testing of the planned Fw 190 C series powered by the DB 603 A engine.

The massive turbo-supercharger, which was installed beneath the wing center section, gave the machine its nickname "Kangaroo." The exhaust pipes, which ran over the wing root and beneath the fuselage, made this prototype approximately 35 kph slower. Maximum boost altitude of this power plant system was roughly 11,500 meters. Anticipated performances at this altitude, where the high-altitude fighter would operate, could not be achieved because the airframes were not sufficiently pressure-tight. During trials, which revealed a performance inferior to the Fw 190 with DB 603 A engine without exhaust turbine at lower altitudes, on 28 May 1943 the Fw 190 V 31 (WNr. 0 056, GH + KU) was lost in a forced landing and as a result was not available to take part in the Ta 152 H test program.

The previously-named prototypes for the Ta 152 H were equipped as follows: power plant: Junkers Jumo 213 E. Forward fuselage extension 775 mm, rear 500 mm. Pressurized cockpit. Aft fuel tank 292 liters, forward 230 liters, sheet metal. Wing as Ta 152 H (area 23.30 m², span 14.44 m). No armament, no GM 1 system).

In addition, a completely new prototype, the Ta 152 V 25 (Wer*knummer* 110 025), was to come from the Sorau prototype shop. The Ta 152 V 25 was thought of as a replacement for the Fw 190 V 33/U1, which had left the test program early on, and in contrast to the previously-mentioned prototypes it had the new wing with fuel tanks and the methanol-water system.

Furthermore, the Focke-Wulf statement of 23 August 1944 anticipated the following equipment status for the Ta 152 H-0 series, production of which was to begin in October 1944 and which differed from the full series Ta 152 H-1 as follows:
1   no fuel tanks in wings
2   no methanol system
3   but with GM 1 system from the first aircraft.

Side view of the Fw 190
V 30/U1, GH + KT, at
Adelheide.

Front and rear views of
the second prototype of
the Ta 152 H, Fw 190
V 30/U1, *Werknummer*
0 055, with the new
high-aspect ratio wing.

That was the plan. Just two months later everything looked quite different. Work on the Ta 152 V 25 prototype was halted as a result of delays. The prototype's already complete wing, which contained four bag tanks, was now to be used on the fifth prototype, the Fw 190 V 32/U1, which was still under construction. The V 32/U1 was thus equipped like the Ta 152 H-1 with GM 1 system in the fuselage, three fuel tanks and one methanol-water tank in the wings, FuG 16 ZY but still no weapons.

Friedrich Schnier's High-Altitude Flight: 13,654 Meters

Oberfeldwebel Friedrich Schnier was a late addition to the Focke-Wulf test program. When the trials unit testing the Ta 154 two-seat night fighter was disbanded, Hans Sander retained him to take the place of Werner Bartsch, who had been severely injured in the crash of the Ta 152 V 9 on 18 April 1944. Schnier continued to fly the Ta 154 in demonstrations and on one occasion he succeeded in out-turning the prototype of the Bf 109 H while flying the Ta 154! The name Schnier is associated with the highest known high-altitude flight in the Ta 152. The last year of the war had already begun, when on 20 January 1945 Obfw. Schnier took off from Langenhagen in the third prototype of the Ta 152 H, the Fw 190 V 29/U1, GH + KS, on a high-altitude flight.

Friedrich Schnier described the preparations and the flight itself: "Before I made this record-setting flight, the highest altitude I had reached was 11,000 meters. But now I was to ascertain the highest altitude that could be reached by the Ta 152 H. At that time our German altimeters registered a maximum altitude of just 12,000 meters. Therefore an Italian altimeter was installed which was good to 14,000 meters. It was checked before and after. The flight proceeded normally at first. I radioed data (speed, altitude, cockpit pressure and temperature) every 1,000 meters. At an altitude

Side view of the third prototype of the Ta 152 H, Fw 190 V 29/U1, *Werknummer* 0 054. The aircraft made its maiden flight on 24 September 1944.

Test pilot Friedrich Schnier reached an altitude of 13 654 meters during an altitude flight in the Ta 152 H, Fw 190 V 29/U1 from Langenhagen on 20 January 1945.

of 10,000 meters I tried to inflate the canopy sealing bladder, but the result was not satisfactory. Because of the leakage the cockpit pressure was not much higher than the outside pressure. Above 10,000 meters I became itchy and had pain in my elbows and knees. I had the feeling that my movements were becoming stiffer. At 12,000 meters I radioed that the standard altimeter was against the stop. I continued to climb slowly and felt that I was higher than ever before. My field of vision grew ever narrower, as in a movie film. The sky was the most beautiful color, ranging from dark blue to black, passing through every shade from dark blue to white on the horizon. Since my right arm no longer obeyed my will, I continued the flight with my left hand. When, some time later, I encountered further difficulties and could climb no further, I decided to head back. I made several more speed tests while descending and radioed the results to ground control in Langenhagen.

It was night when I landed there, and I found the technicians, who had monitored my entire flight by radio tensely awaiting my arrival. All were anxious to see what evaluation of the barograph would reveal.

The data could be read from the strip chart. It revealed that I had reached an altitude of 13,654 meters."

Series of Accidents during Testing

The first test report on the Ta 152 H with Jumo 213 E of 30 January 1945 incorporated all test results achieved by all of the completed prototypes and is repeated here in abbreviated form. The following aircraft were delivered by the prototype shop in Adelheide by the start of production of the Ta 152 H-1:

1   Fw 190 V 33/U1
    Werknummer 0 058, Ta 152 H
    Registration: GH + KW
    First flight on 13 July 1944

2   Fw 190 V 30/U1
    Werknummer 0 055, Ta 152 H
    Registration: GH + KT
    First flight on 6 August 1944

3   Fw 190 V 29/U1
    Werknummer 0 054, Ta 152 H
    Registration: GH + KS
    First flight on 24 September 1944

4   Fw 190 V 18/U2
    Werknummer 0 040, Ta 152 H
    Registration: CF + OY
    First flight on 19 November 1944

1   Ta 152 H 0 058 was lost to the test program after just 36 minutes flying time. During the ferry flight from Adelheide to Langenhagen it made a crash landing near Vechta and sustained 70% damage. While 058's landing flaps and undercarriage functioned perfectly on the test stand prior to its maiden flight, in flight the right undercarriage leg could not be locked in position because the movable wheel well cover jammed against the fixed landing gear door. The doors were readjusted for the second flight, but nevertheless the right undercarriage leg refused to lock. No closer examination of the cause could be made, as the aircraft crashed during the ferry flight.

2   Ta 152 H 0 055 was lost to the test program after 10 hours, 3 minutes flying time when it crashed on 23 August 1944. The aircraft's Jumo 213 E caused problems during testing. On every high-altitude flight the third stage of the supercharger refused to engage and above 9,000 meters there was a drop in fuel pressure caused by fuel tank pumps not suitable for high-altitude operation. The Fw 190 V 30/U1 was flown by the Rechlin Test Station for the first time 19 August 1944. During a high-altitude flight 23 August the Jumo 213 E caught fire. While the fire spread no further, GH + KT crashed in a turn during the landing procedure and was 100% destroyed. Although it is not mentioned in the report, *Flugkapitän* Alfred Thomas lost his life in the crash of the V 30/U1. Hans Sander described the accident as follows: "Following an engine fire at high altitude, the V 30/U1 crashed while on approach to land at Adelheide. That is all we know, as there was no radio contact."

3   Ta 152 H 0 054 was the first Ta 152 H which could be extensively tested over a longer period. On 27 September 1944 the Rechlin Test Center evaluated its flight characteristics at an all-up weight of 4,200 kg and made the following assessment:

Assessment Summary

1   Trim changes around the pitch axis as a result of lowering landing flaps bearable.
2   Stall behavior is not comfortable, but can be seen as acceptable.
3   Stability about the vertical axis weak. Aircraft has a certain tendency to skid.
4   The aircraft is stable about the pitch axis at the center of gravity positions (to 0.665) flown to date.

When production began it became apparent that carriage of the 300-liter drop tank on the ETC 500 rack worsened the aircraft's already poor stability situation about the vertical axis. Use of the ETC 504, which was located 300 mm further aft, eliminated the reduction in stability.

The maximum altitude achieved by GH + KS was 13,654 meters on 20 January

1945. It was discovered that the over-pressure valve was set at 9,750 meters instead of 8,000. It was also found that ice formed on the canopy and windscreen. There were also problems with the Jumo 213 E which were identified as supercharger surging. An attempt to alleviate the problem through the use of an enlarged vent line with a cross-section of 15 cm² proved unsatisfactory. GH + KS had logged a total of 20 hours, 13 minutes in the air by the time production began. The V 29/U1 was unavailable for testing from 2 to 27 November 1944 as a result of engine failure (several cylinders had zero compression). During a performance flight on 31 January 1945 the V 29/U1 reached a maximum speed of 708 kph at an altitude of 10,800 meters.

The altitude flight on 20 January 1945 (pilot Schnier) revealed a climb rate of 16.8 m/sec at ground level, dropping to 16 m/sec by 2,500 meters. The maximum altitude reached was 13,650 meters. These performances were achieved without GM 1 or MW 50.

4 Ta 152 H 0040 emerged from the prototype shop just as series production was beginning and therefore could not be used in testing. It was ferried from Adelheide to Langenhagen on 19 November 1944 and from 21 to 25 November 1944 was based at Cottbus for familiarizing pilots with the new Ta 152 H. On 23 December 1944 the V 18/U1 was slightly damaged in a takeoff crash, and during the subsequent repairs 0040 received the new wooden tail planned for Ta 152 H production aircraft.

Taken altogether, the prototypes flew a total of just 30 hours, 52 minutes prior to the start of production.

Flight times by 30 January 1945
Ta 152 H, Werknummer 0 058
0 hours 36 minutes
Ta 152 H, Werknummer 0 055
10 hours 3 minutes

Side view of the Fw 190 V 32/U1, GH + KV, seen here during trials as the so-called "quick solution" for the high-altitude fighter with turbosupercharger removed and DB 603 engine.

The Fw 190 V 32/U1 was supposed to be the first prototype of the Ta 152 H equipped with the new wing.

Ta 152 H, *Werknummer* 0 054
36 hours 1 minute
Ta 152 H, *Werknummer* 0 040
5 hours 2 minutes
Total flying time
49 hours 42 minutes

Summary
The entire flight test program was set back by the early loss of the Fw 190 V 33/U1 and V 30/U1. As a result, important tests could not be carried out or were scaled back. The important fifth prototype, the Fw 190 V 32/U1 (*Werknummer* 0 057, GH + KV), which was built to the standard of the planned production Ta 152 H-1, was unable to join the test program before production began or even before 31 January 1945. There was therefore no information on the type's flight characteristics with wing tanks or the MW 50 and GM 1 systems. Unfortunately, serious problems with the installation of these systems were encountered during production of the Ta 152 H-1 series, resulting in the GM 1 system being dropped from production aircraft. In spite of this stability problem there was never any doubt that the Ta 152 H would enter *Luftwaffe* service. Overall it can be said that there were too few prototypes and that these began trials much too late for an aircraft which was to have such an important role; one of the main reasons for this was the massive withdrawal of specialist personnel from the prototype construction unit.

The Variants of the Ta 152

Production of the Ta 152 H began with the Ta 152 H-0 variant. The Ta 152 H-0 was powered by a Junkers Jumo 213 E engine which developed 1,730 H.P. for takeoff at 3,250 rpm. The weapons system consisted of one 30-mm MK 108 engine-mounted cannon with electro-pneumatic cocking and electric firing. The MK 108 fired through the hollow propeller spinner and had 90 rounds of ammunition. The aircraft also carried two MG 151/20 cannon in the

wing roots with 175 rounds per gun. The Ta 152 H-0 still had the one-piece wing with a span of 14.4 meters; however, it had no wing tanks, which limited its range. The Ta 152 H-0 could carry a total of 595 liters of fuel in the fuselage, divided between a forward tank with a capacity of 233 liters and a rear tank with 362 liters. As well, a 300-liter drop tank could be carried on a Type 503 A-1 rack. Although the tank compartment aft of Bulkhead 8 was designed for the GM 1 tank, the eighteen Ta 152 H-0s in production were equipped with a 115-liter auxiliary tank there. No MW 50 system was anticipated for the H-0 variant.

The next variant, the Ta 152 H-1, differed from the H-0 series in having the two-piece wing which was equipped with six unprotected bag tanks. The left inner bag tank was chosen to contain 70 liters of MW 50, while the remaining five tanks raised the Ta 152 H-1's total fuel by 400 liters.

Maintenance panels in the underside of the wing provided access to the tanks. In the fuselage, the 85-liter GM 1 tank was again installed aft of Bulkhead 8. From the first aircraft the Ta 152 H-1 was equipped with Rüstsatz 8, for bad-weather operations; it consisted of the FuG 125 Hermine, the LGW K23 automatic pilot and heated windscreen. The enormous increase in the Ta 152 H-1/R11's fuel capacity led to stability problems, which were alleviated by the temporary measure of eliminating the GM 1 system. The GM 1 system was also not used on the Ta 152 H-1/R21, which received the MW 50 high-pressure system. Not until the H-1/R31, which had ballast installed in the engine and a limit of 280 liters placed on the rear fuel tank, was full use of the GM 1 system restored.

Further changes intended to address the stability problem, such as a modified wing-fuselage fairing and an increase in tail surface area, could not be introduced into Ta 152 series production.

The Ta 152 H's radio equipment consisted of the VHF FuG 16 ZY and the FuG

Fw 190 V 18/U2, *Werknummer* 0 040, was the fourth Ta 152 H prototype and did not join the test program until 19 November 1944. Here it is seen as the "*Höhenjäger* II" prototype with exhaust-driven turbo-supercharger and DB 603 S.

The Fw 190 V 18/U2, CF + OY, was parked at Reinsehlen on 6/04/ 1945 and was blown up a short time later.

25 IFF set. The Ta 152 H-2 differed from the H-1 only in having the new FuG 15 in place of the FuG 16 ZY. Demand for this series ended on 15 December 1944, however, because the necessary ground organization could not be converted in time. Nevertheless, one Ta 152 H-2 was delivered in April 1945.[1]

During production of the Ta 152 H there was a change in power plants from the Jumo 213 to the Jumo 213 E-1. The Jumo 213 E-1 had a reinforced transmission. Experience with the first 200 Jumo 213s delivered revealed that the use of emergency power resulted in transmission failures. As well the problem of supercharger surging had not been entirely cured, as a result of which a bleed valve had to be installed on the Jumo 213 E-1. Deliveries of Jumo 213 E-1 engines with the MW 50 high-pressure system, which resulted in a considerable

improvement in MW 50 operation, were supposed to begin on 29 April 1945. The Jumo 213 E-1 developed 2,100 H.P. for takeoff and 1,600 H.P. at an altitude of 8,200 m. From about 1 July 1945 the Ta 152 H was to be powered by the Jumo 213 EB, whose output was 200 H.P. greater than that of the Jumo 213 E. The long-range goal was the introduction into service in November 1945 of the Jumo 213 I engine, which offered 2,700 H.P for takeoff and 1,900 H.P. at 10,000 meters. This output was equivalent to a jet engine with 1,000 kg of thrust at the same altitude. High-altitude fighter was not the only role anticipated for the Ta 152 H by the RLM; a high-altitude reconnaissance version was also called for. It received the designation Ta 152 H-10 and was based on the Ta 152 H-0. The Ta 152 H-10 and H-11 corresponded to the Ta 152 H-1 and H-2. None of the reconnaissance aircraft were built.

[1] Aircraft allocation OKL/ Gen.Qu.(6 Abt.IIIC): 2/ 04/45 one Ta 152 H-2 to Luftflotte Reich.

Another interesting view of the blown-up V 18/U2, here from the tail looking forward.

The Variants of the Ta 152 H

Designation	Engine Tank	Wing	GM1	MW 50	Forward	Rear	Total	Remarks
H-0	Jumo 213 E	—	—[15]	—	233 l	362 l	595 l	Pre-production series
H-0/R11	Jumo 213 E	—	85 l	—	233 l	362 l	595 l	Pre-production bad weather fighter
H-1	Jumo 213 E/E-1	400 l	85 l	70 l	233 l	362 l	995 l	Until 9/3/45 MW 50 low-pressure system
H-1/R11	Jumo 213 E/E-1	400 l	85 l	70 l	233 l	362 l	995 l	Until 9/3/45 MW 50 low-pressure system
H-1/R11	Jumo 213 E-1/EB	400 l	off	70 l	233 l	362 l	995 l	From 9/3/45
H-1/R21	Jumo 213 E-1/EB	400 l	off	70 l	233 l	362 l	995 l	MW 50 high-pressure system installed
H-1/R31	Jumo 213 E	400 l	85 l	70 l	233 l	362 l	995 l	MW 50 high-pressure system installed, ballast in engine
H-2	Jumo 213 E	400 l	85 l	70 l	233 l	362 l	995 l	FuG 15 instead of FuG 16 ZY, canceled 15/12/44
H-2/R11	Jumo 213 E	400 l	85 l	70 l	233 l	362 l	995 l	FuG 15 instead of FuG 16 ZY, canceled 15/12/44
H-10	Jumo 213 E	—	—	—	233 l	362 l	595 l	High-altitude reconnaissance aircraft based on Ta 152 H-0
H-0 H-11 (planned)	Jumo 213 E/E-1	400 l	85 l	70 l	233 l	362 l	595 l	High-altitude reconnaissance aircraft based on Ta 152 H-1
H-12	Jumo 213 E/E-1	400 l	85 l	70 l	233 l	362 l	595 l	High-altitude reconnaissance aircraft based on Ta 152 H-2

The following power plant changes were supposed to take effect during production.[16]

Until approx. 1/3/45	Jumo 213 E engine with MW 50 low-pressure system (BNE system). For these engines emergency power had to be shut off in third stage. Furthermore, vent lines were to be installed to eliminate supercharger surging
From approx. 1/3/45	deliveries were supposed to begin of Jumo 213 E-1 engines cleared for use of emergency power with a bleed valve to eliminate supercharger surging.
From roughly 25/4/1945 1/6/1945	these engines were to be replaced by the Jumo 213 E-1 engine with MW 50 high-pressure system. A further change was to follow from approximately from the Jumo 213 E-1 with high-pressure system to the Jumo 213 EB with intercooler and Jumo cooler head.
Rüstsatz R11:	Additional equipment for the bad-weather fighter role. It was supposed to be used from the first Ta 152 H-0 and the first Ta 152 H-1. It consisted of the FuG 125 Hermine, the LGW K23 automatic pilot and heated windscreen.

Changes to series designations were necessary from 9 March 1945 as a result of stability problems.

[1] Focke-Wulf notification of power plant and stability changes Ta 152 H-1/R11 series dated 9 March 1945.
[2] Installation of 115-liter B4 fuel tank on 18 aircraft.

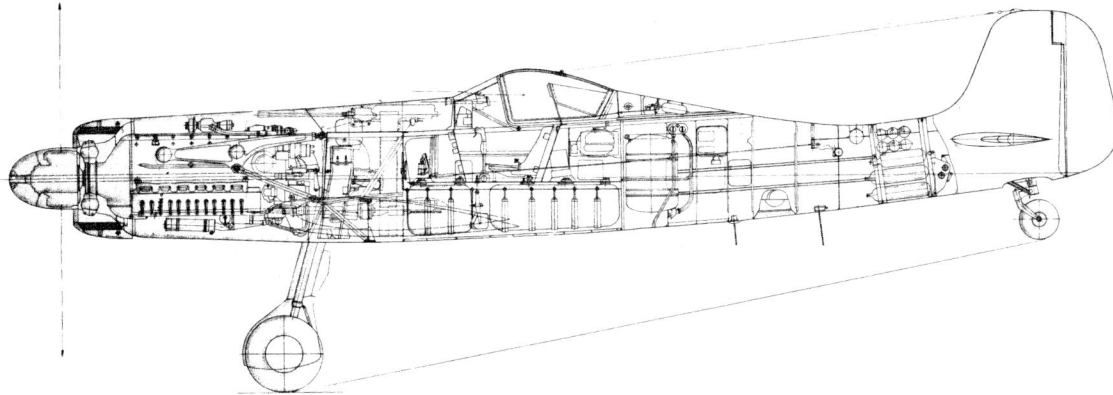

Begleitjäger Ta 152 H (Längsschnitt)

Triebwerk: Einheitstriebwerk Jumo 213 E
mit GM1 Anlage oder Reichweitenbehälter

1.5.44

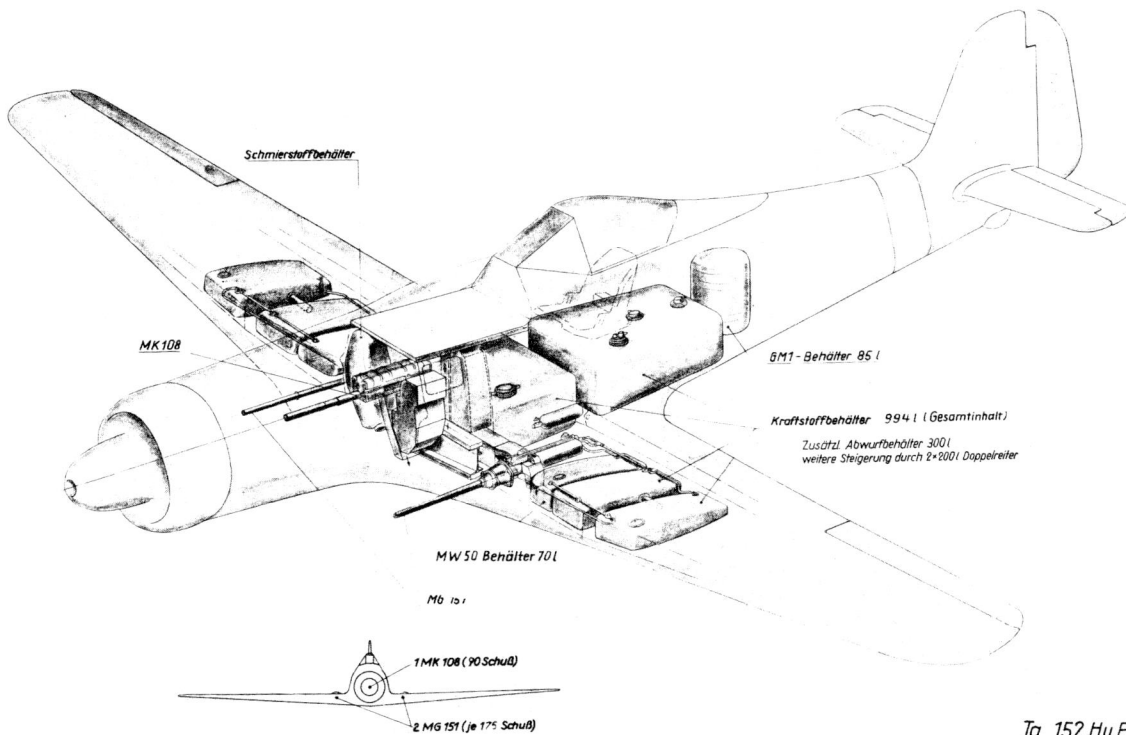

Schmierstoffbehälter

MK 108

GM1 - Behälter 85 l

Kraftstoffbehälter 994 l (Gesamtinhalt)

Zusätzl. Abwurfbehälter 300 l
weitere Steigerung durch 2×200 l Doppelreiter

MW 50 Behälter 70 l

MG 151

1 MK 108 (90 Schuß)

2 MG 151 (je 175 Schuß)

Ta 152 Hu.E

Waffen- und Behälteranlage

87

Hydraulische Fahrwerksanlage der Ta 152 H

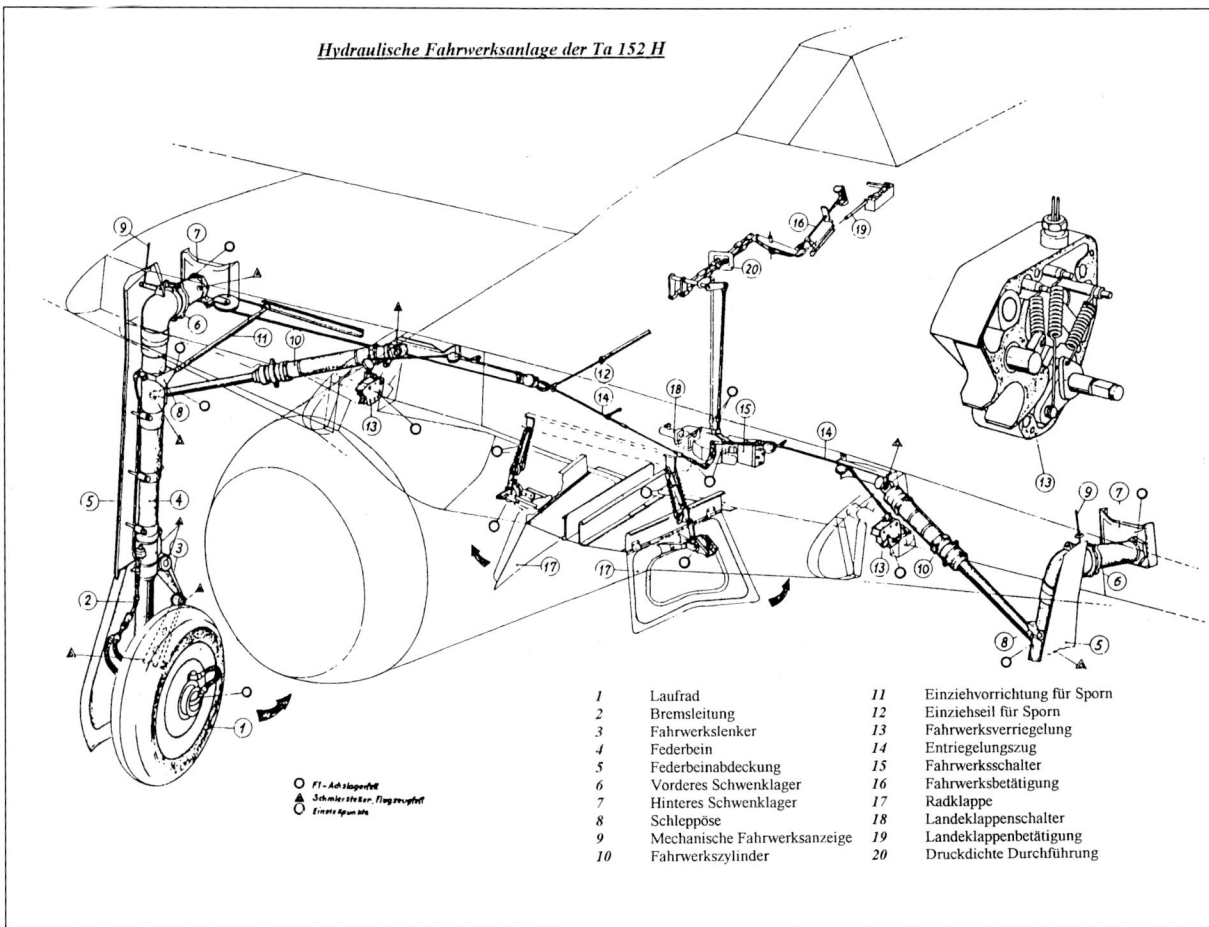

1	Laufrad	11	Einziehvorrichtung für Sporn
2	Bremsleitung	12	Einziehseil für Sporn
3	Fahrwerkslenker	13	Fahrwerksverriegelung
4	Federbein	14	Entriegelungszug
5	Federbeinabdeckung	15	Fahrwerksschalter
6	Vorderes Schwenklager	16	Fahrwerksbetätigung
7	Hinteres Schwenklager	17	Radklappe
8	Schleppöse	18	Landeklappenschalter
9	Mechanische Fahrwerksanzeige	19	Landeklappenbetätigung
10	Fahrwerkszylinder	20	Druckdichte Durchführung

○ Fl-Achslagerteil
▲ Schmier Stelle bei Flugzeugteil
○ Einzelteil-Spornteile

Focke-Wulf Flugzeugbau G. m. b. H.   Nr. 26 a

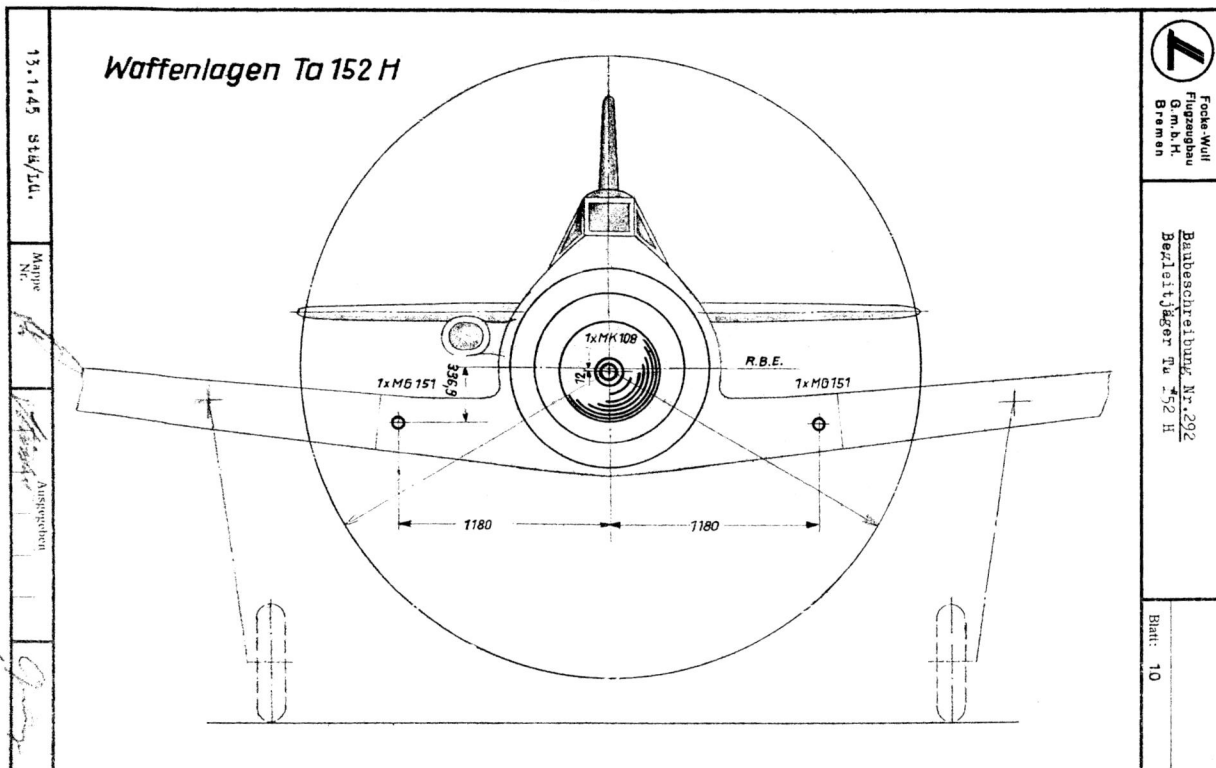

Waffenlagen Ta 152 H

13.1.45   Stb/Ldu.

Mappe Nr.

Ausgegeben

1 x MK 108
1 x MG 151
1 x MG 151
R.B.E.
336,9

1180   1180

Focke-Wulf
Flugzeugbau
G.m.b.H.
Bremen

Baubeschreibung Nr. 292
Begleitjäger Ta 152 H

Blatt: 10

88

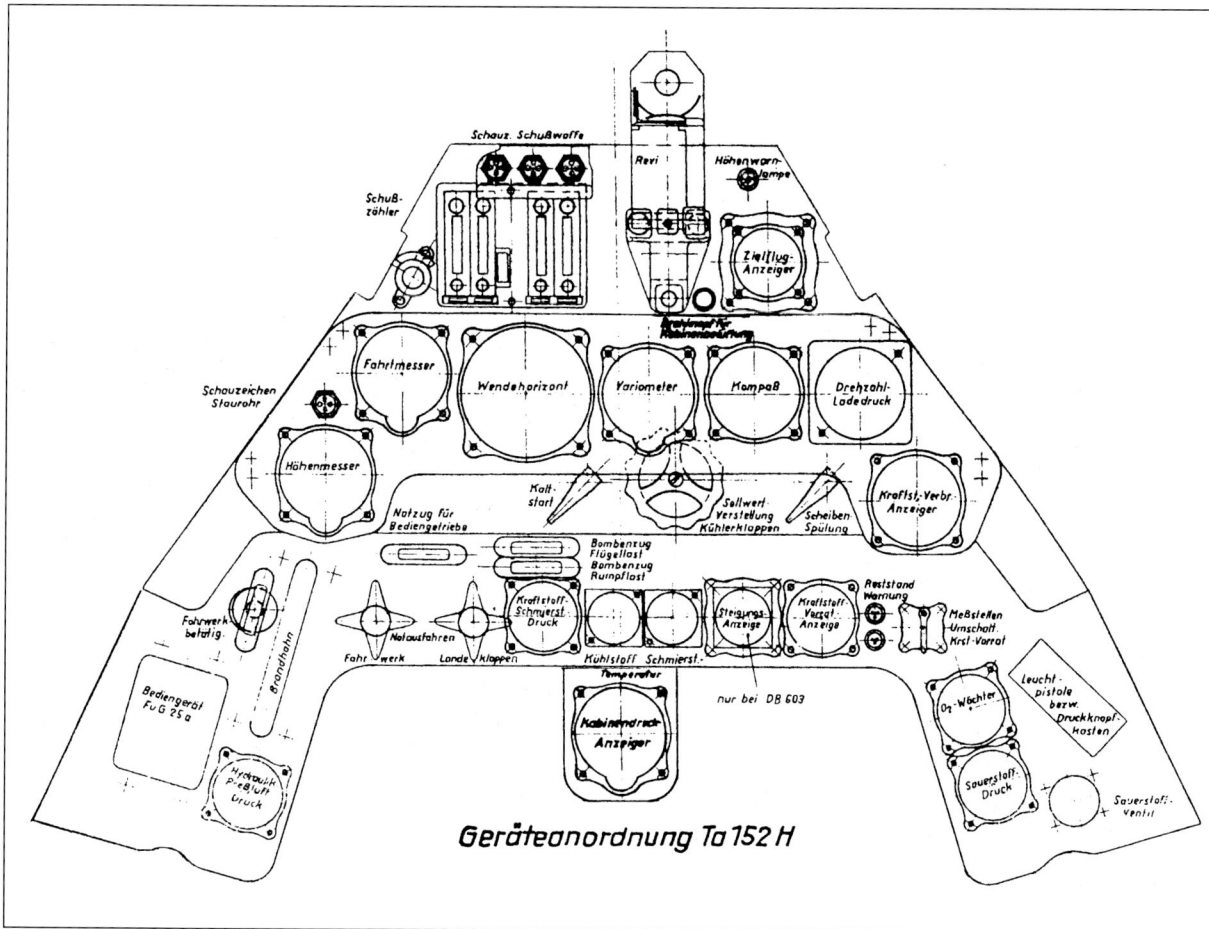

Geräteanordnung Ta 152 H

Schauz. Schußwaffe
Schuß-zähler
Revi
Höhenwarn-lampe
Zielflug-Anzeiger
Schauzeichen Staurohr
Fahrtmesser
Wendehorizont
Variometer
Kompaß
Drehzahl-Ladedruck
Höhenmesser
Natzug für Bediengetriebe
Kalt-start
Sollwert-Verstellung Kühlerklappen
Scheiben Spülung
Kraftst.-Verbr. Anzeiger
Fahrwerk betätig.
Brandhahn
Notausfahren
Fahrwerk
Lande-klappen
Kraftstoff-Schmierst. Druck
Kühlstoff Schmierst.-Temperatur
Bombenzug Flügellast Bombenzug Rumpflast
Steigungs-Anzeige
Kraftstoff-Vorrat Anzeige
Reststand Warnung
Meßstellen-Umschalt. Krst.-Vorrat
Leucht-pistole bezw. Druckknopf-kasten
Bediengerät FuG 25 a
Hydraulik Preßluft-Druck
Kabinendruck-Anzeiger
nur bei DB 603
O₂-Wächter
Sauerstoff-Druck
Sauerstoff-ventil

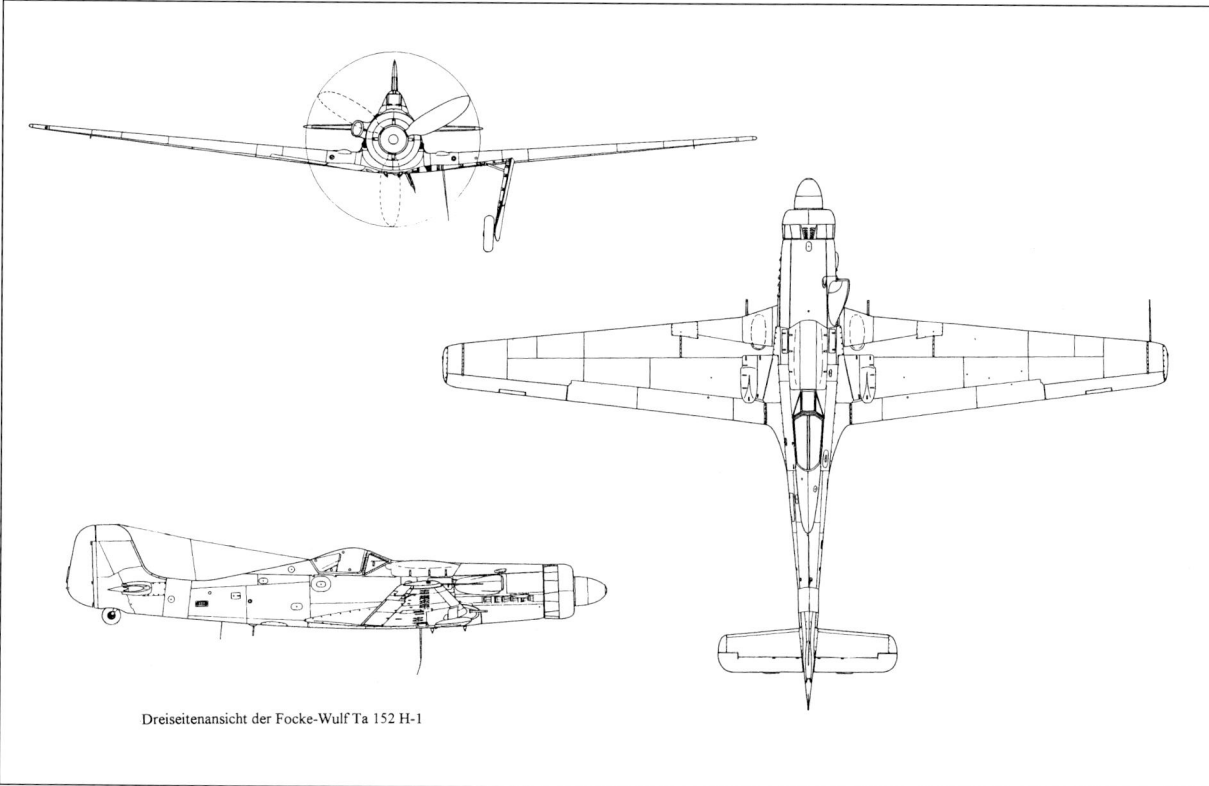

Dreiseitenansicht der Focke-Wulf Ta 152 H-1

89

G e w i c h t s a u f s t e l l u n g

Benennung	Ta 152 H-0	Ta 152 H-1
Rumpfwerk	412 kg	412 kg
Fahrwerk	245 kg	245 kg
Leitwerk (Metall)	136 kg	136 kg
Steuerwerk	35 kg	33 kg
Tragwerk	629 kg	654 kg
Triebwerk vor Brandschott	1822 kg	1822 kg
Triebwerk in der Zelle	170 kg	248 kg
Normale Ausrüstung	224 kg	247 kg
Zweckausrüstung	233 kg	233 kg
Ballast	14 kg	1 kg
Rüstgewicht	3920 kg	4031 kg
Flugzeugführer	100 kg	100 kg
Kraftstoff im Rumpf vorn	172 kg	172 kg
" " " ninten	268 kg	268 kg
Kraftstoff im Rumpfzusatzbehälter 115 l	85 kg	–
Kraftstoff in 4 Flügelbehältern 400 l	–	296 kg
GM1 im Rumpf hinten 85 l	–	104 kg
MW 50 im Flügel l.innen 70 l	–	64 kg
Schmierstoff	55 kg	55 kg
Munition 2 MG 151 je 175 Schuß	77 kg	77 kg
" 1 MK 108 m. 90 Schuß	50 kg	50 kg
Zuladung	807 kg	1186 kg
Normal—F l u g g e w i c h t	4727 kg	5217 kg

Focke-Wulf Flugzeugbau G.m.b.H. Nr. 26a

The Ta 152 H in Service

Although quantity production of the Ta 152 H began only two months after the Fw 190 D-9 in November 1944, the first Ta 152 H did not reach the *Luftwaffe* until 27 January 1945. The difficult production conditions and problems with deliveries of components prevented the Ta 152 H from being built in large numbers. The rapid approach of the front in the east very quickly brought production to an end in Cottbus. The few Ta 152 Hs to see action made a very good impression on the pilots who flew them. Obfw. Willi Reschke declared: "I would have been pleased if I had always had a machine such as this with its performance and handling qualities during my missions and air combats. I still consider the Ta 152 to have been first class for the conditions and demands at that time." It is not known whether later Ta 152 H-1s with MW 50 or GM 1 power boosting were flown in action, and former pilots are also unable to confirm this. While the first Ta 152 H-0s did not have these systems, thanks to their low weight they were fully capable of performances in the same class as that of the P-51 D Mustang. The following tactic was successfully used: if the Ta 152 H pilot survived the first attack by the Mustang he could use the tighter turning circle of his

The third production Ta 152 H-0, seen here in front of the factory hangars in Cottbus, was the first to arrive at Rechlin, on 11 December 1944.

Side view of Ta 152 H *Werknummer* 150 003 with 300-liter drop tank on the new Ta tank carrier.

Ta 152 H-0 CW + CC was the first to be equipped with MW 50 injection.

aircraft to reach firing position, turning the hunter into the hunted. At least one Mustang was credited to III/JG 301. The pilots of the Tank also did not have to fear P-47 Thunderbolts or Hawker Tempests, as several victories prove.

The Ta 152 H never served in the role it was designed for, namely the interception of high-flying Allied bombers and recon-naissance aircraft. In spite of this, several pilots of III/JG 301 made high-altitude flights; on one occasion Obfw. Willi Reschke reached 12,500 meters. The purpose of these flights was not to fly as high as possible, but to investigate the handling qualities of the Ta 152 H at those altitudes.

In spite of repeated claims to the contrary, the Ta 152 H was never used to protect Me 262 fighters during takeoff and landing.

Report by the Focke-Wulf Technical Field Team (TAT)[1]

On 14 February 1945 Focke-Wulf's technical field team released a report on trials with the Ta 152 H-0 by III/JG 301. The report was five pages long and was divided into the following sections:

Assessment of the Ta 152 H-0 by the unit:

a) Flight characteristics
b) Maintenance load
c) Criticisms
1) Undercarriage
2) Wing
3) Power plant
4) Compressed air system supply situation

The complete report is reproduced here.

Assessment of the Ta 152 H-0 by the Unit

III/JG 301 unanimously gave the Ta 152 H-0 the best evaluation that the undersigned (Herr Martin of the TAT) ever received from a front-line unit concerning a Focke-Wulf product. The aircraft's excellent handling qualities in the turn came in for particular praise. The extent of the complaints received to date are below the level to be expected for a new type. Apart from undercarriage hydraulics, they do not have the character of fundamental errors which might jeopardize the immediate front-line use of the aircraft. Where elimination of these shortcomings was immediately necessary in the interests of maximum readiness, the *Gruppe* helped itself. It is, however, expected that these problems will have been eliminated from the next batch of aircraft to be delivered, especially since the necessary changes are not extensive. In anticipation of the fastest possible elimination of these shortcomings the *Gruppe* did not forward the complaints to higher offices, instead it only described the good overall impression with the comment that those shortcomings detected were being overcome in direct cooperation with the manufacturer. The most important shortcomings to be addressed are detailed below.

Flight Characteristics

Compared to the Fw 190 A-8, the Ta 152 H-0 is capable of tighter turns with less tendency to fall off into a spin, and this only happens at lower airspeeds (approx. 250 kph). Aircraft which enter a spin in this way can easily be recovered after about 500 to 600 meters by pushing the nose down. Naturally the larger wing has reduced maneuverability somewhat, but this is in no way seen as a disadvantage. The *Kommandeur* claims that he began to black out while turning, something that never happened to him while flying the Fw 190. In a dogfight with an A-8 the latter, flown by a very good pilot, was easily outturned by the H-0, which was flown by a pilot with only two flights behind him in the Ta. All of these characteristics have so far mainly been tested from ground level to about 3,000 meters. The *Kommandeur* has achieved climb performances of eight minutes to 7,000 meters and fifteen minutes to 11,000 meters, and he pointed out that he did not use the most favorable climbing speed. No dives have been carried out yet, or only at speeds below those recommended as safe in the pilot's notes. This caution is attributable to statements by the Rechlin Test Station, which claims to have noted serious instability in diving flight above 600 kph. Instability about the vertical and elevation axes was stressed by Rechlin as an especially negative characteristic of the Ta 152 H-0, requiring constant retrimming in the climb and constant monitoring of the turn and bank indicator while turning. So far the *Gruppe* has made no complaints about these characteristics, which is perhaps due to the fact that only air combat practice flights have been carried out so far and not gunnery flights, which require precise aiming and which would have revealed the instability if it exists to the degree described by Rechlin.

Speed at low level has been measured by several pilots with the following results, whereby the indicated airspeed was compared with that of an A-8 as a check.

Take off and landing runs are very short and according to the *Kommandeur* permit the use of fields which cannot be identified as fighter bases from above.

[1] TAT report dated 14/2/1945

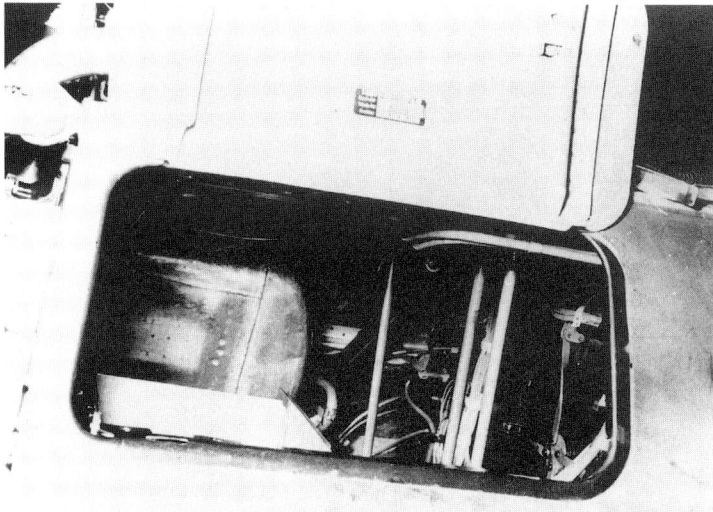

Open servicing hatch behind the firewall of *Werknummer* 150 003. The hatch provided access to the MK 108 and the oil tank filler point.

Maintenance Load

The technical officer assessed the maintenance load of the Ta as less than that of the Fw 190; however, it must be taken into consideration that there is still no maintenance manual available with the necessary tips and the aircraft are not yet equipped with pressurized cockpits. Furthermore, experience gained with high-altitude time has yet to be evaluated.

Criticisms

Undercarriage

a)   In flight the undercarriage does not retract in time. The causes are unknown. After the tailwheel had to be left down (retraction mechanism did not function properly), it was found that the tailwheel could be retracted with no problems after a takeoff with 20 degrees of flap if the retract command was given while simultaneously adopting a slightly nose-down attitude at about 200 kph (therefore no double retract command). This method is only a stopgap, however, which must be eliminated as quickly as possible through reliable operation of the hydraulics, whose load must, if

necessary, be lightened by locking the wheel well doors in the down position. The *Gruppe* will not want to accept a perceptible loss of airspeed. It should be possible to compensate for this, since all of the speeds previously cited were achieved with the tailwheel extended.

b)   The tailwheel retraction mechanism's roller hops from the guide track on the spar. The cause could not be determined with certainty. In several cases the roller may have come out of the guide track as a result of incorrect positioning, while in two instances badly widened tracks were found, in which the roller, operating with a great deal of play, probably edged over.

c)   The return spring for the locking lever is too weak. Especially in operations from softened airfields, mud enters the incompletely encapsulated roller, making the retraction cycle even more difficult. The return spring must be strengthened. As a stopgap measure the *Gruppe* has installed an additional spring.

d)   During takeoffs and landings from soggy airfields a great deal of water and sand or mud is sprayed into the undercarriage wells. From there water, for the most part, makes its way through the MG 151 mounts into the spar for the hydraulic lines as well as through the heating tube junction for outboard weapons in the MG 151 weapons bay. On one occasion forty mm of water was still present there approximately fifteen minutes after landing, since there are no drain holes. This poses a danger to the cannon breeches, the ADSK and the antenna coupler for the Y antenna. Furthermore, mainly earth was thrown against the engine-mounted cannon, clogging up the breech, creating the danger of unintentional firing. The covers planned to date, for example the blast tubes around the MG 151s

Altitude:	2,100	2,400	2,700	3,100
Kph:	390-430	450-470	490-510	550

94

and the engine compartment cover, absolutely must be installed. The *Gruppe* has helped itself in the short term by installing appropriate covers and drainage holes (6-mm) in the cannon bays.

e)   Unlike the outer bushing, the inner bushing of the upper coupling of the undercarriage actuator is not secured by set screws. In one case this bushing fell out and was responsible for the failure of the actuator to lock in the extended position.

f)   On all aircraft a heavy buildup of rust is noticeable on the bare pistons of the actuators after a few hours. An attempt must be made to protect against rust through the installation of oil or graphite rings at the piston exit and the application of grease. In any case, daily degreasing is to be added to the servicing manual.

g)   In a total of three instances leakage of hydraulic fluid was detected from the threads of the lower attachment eye on the

Detail view of the wheel well door retraction mechanism of the Ta 152 H.

Remains of a Ta 152 H wheel well door which was recently discovered in Neustadt-Glewe.

actuator piston. The actuators had to be replaced.

h)   Involuntary retraction of the undercarriage occurred on two occasions, once during engine start-up, in which the prescribed preliminary pulling of the undercarriage handle was omitted. It proved impossible to ascertain who could have set the undercarriage switch to RETRACT in the brief interval during the change of pilots. In the other case touchdown probably occurred before the actuator had locked in the extended position. A check revealed that the hydraulic switch had not yet sprung back to the neutral position, therefore had not been switched off yet.

Wing

a)   Loose rivets. As previously reported in the second report dated 17 January 1945, loose rivets appear on all Ta 152 H-0 aircraft beneath the leading edge of the wing on the forward rim of the wheel cut-out in the wing.

Power Plant

a)   Evaporation occurs, especially during takeoff, as a result of which glycol is sprayed onto the windscreen because of the unfavorable location of the line. The windscreen washer, which uses gasoline, is unable to remove the glycol. The evaporation line must be repositioned.

b)   In two instances already the fairing over the oil cooler detached in flight and fell into the propeller. Since the destroyed parts were no longer available, it could not be determined whether this was the result of failure of the mountings or improperly secured bolts.

Compressed Air System

a)   Because of poor rubber material, valve Hn 2 prevented the air bottles from filling after emergency activation of the undercarriage and in many cases leaked, allowing compressed air to leak at the outboard attachment. In some cases the rubber material stuck to the body of the valve or was hard and porous.

Supply Situation

The procurement of spares for the ten Ta 152 H-0 under test by III/JG 301 at Altenow has so far presented no serious difficulties. The demand for spare parts is moderate and can be met by Fw Cottbus. However, this source will soon be exhausted for Altenow. Elbag Camp 256 in Tetschen-Bodenbach, to which the *Gruppe* had been referred, was still completely unstocked on 13 February 1945. The *Gruppe* was not informed that the delivery of Ta 152 spares had been redirected to Elbag Camp 288 at Berlin-Tempelhof. Apart from that, in terms of transport (including couriers) Berlin is not seen as an ideal base. The question of whether

the wrecked aircraft still at Neuhausen could be cannibalized in urgent cases (provided that the major components have not already been earmarked for the production program) was therefore discussed once again with management and construction supervision in Cottbus. Construction supervision agreed that the *Gruppe* could salvage the necessary material in Neuhausen after submitting a requisition for approval.

Remarks

So far the wing of the Ta 152 H-0 has proved less prone to damage in belly landings or landings with one undercarriage leg retracted. No wing damage has occurred in any of the cases so far of the wing contacting the ground as a result of improper landings.

> Bremen, 19 February 1945
> Technical Field Team TAT
> Ma/La.
> (Martin)

Report by Focke-Wulf's Rechlin Office

On 16 March 1945 the Rechlin office of Focke-Wulf wrote a report on the testing of the Ta 152, probably for the last time. This report is repeated here in its entirety. It clearly shows the problems in introducing the Ta 152 and describes the technical detail problems encountered with the Ta 152 H.

Subject 1: Ta 152 Testing

1.   Wooden Tail

Because of the fuel shortage only one flight has been made with the wooden tail installed on WNr. 150 003. Except for the following, no complaints were made about the tail. An error in installation of the cable return pulley for the tailwheel left the tailwheel hanging when the undercarriage was lowered. As a result a landing had to be carried out on the emergency skid, causing damage to the tail at the tailwheel axle mounting points. Flight safety was not endangered as a result. Repairs will be carried

The Ta 152 H-0 was later converted to the new wooden tail at Rechlin and underwent limited testing with it.

out as soon as the machine enters the workshop for other reasons.

At present the second tail is being fitted to WNr. 150 010, after which it will be tested. Unfortunately no extensive testing of the wooden tail will be possible at Rechlin in the near future, because flying is severely restricted by the fuel shortage mentioned above.

2   Undercarriage Testing

At present accumulators of about 40 cm³ capacity are being fitted to the undercarriage hydraulics switches of all aircraft at Rechlin, further the undercarriage locking has been set so that there is a gap of approximately 4 mm between the head of the downlock latch and the roller on the undercarriage leg, finally eliminating the contact between mainwheels and wheel well doors. Six machines have already been modified and adjusted. A concluding assessment of this measure cannot be given, because only two of the machines have been flown, one time each. Everything went smoothly with one machine, in the case of the other the right undercarriage leg dropped shortly before locking but locked in place on the second attempt. Unfortunately airspeed was not observed. All six machines locked perfectly on the test stand at approximately 60 to 70 atmospheres.

3   New Complaints

The pressure reducer on the sliding hood of WNr. 150 011 was set at 12 atmospheres instead of 4, as a result of which the sealing tube between the recess in the hood and the fuselage came out and burst. On two aircraft the radiator fairing tore off and flew away. More secure fastening is required.

The cross-section of the sealing tube recess on the sliding hood is very irregular, as a result of which the tube is badly compressed in places and cannot emerge (also danger of damage).

4   General

On orders of the OKL part of the Rechlin testing operation is being transferred to Lechfeld in Bavaria, including all jet-powered aircraft: Me 262, Ar 234 and He 162, further all departments involved with jet engines or their accessories.

The focus of the Rechlin operation remains at Rechlin as before. Efforts at Rechlin will concentrate on testing the Ta 152, Fw 190, Do 335 and variants of the remaining, still up-to-date types.

5   English Fighter Aircraft Hawker Tempest

Several days ago a new fighter in service with the English air force, the Tempest, was ferried to Rechlin for testing. Although no

performance measurements have been made, performance may be somewhat less than the claimed 690 kph at maximum boost altitude of 6,000 meters with emergency power using the Napier Sabre II A engine. Rechlin estimates performance with short-term emergency boost as 560 to 570 kph at ground level and 670 to 680 kph at the maximum boost altitude of 6,000 m.

Rechlin estimates that installation of the Sabre VI engine, which has yet to reach the front, will result in speeds of about 620 kph at ground level and 730 kph at maximum boost altitude of 8,000 meters using short-term emergency boost.

The Sabre II engine has an emergency output of 2,230 H.P. at 3,700 rpm and 1.66 atmospheres boost pressure. Rechlin made the following assessment of its flight characteristics:

a   Longitudinal stability (static and dynamic) generally good with high stick forces.

b   Dynamic directional stability. Oscillation period 2 1/2 sec. at 2,000 m at speed of 450 kph (oscillation moderates after 4 to 5 cycles).

c   Rudder forces moderate.

d   Aileron forces increase sharply above dynamic pressure and especially at maximum deflection. Stall behavior probably poor, however not yet confirmed since throttle lever fractured on this flight and stall behavior could not be observed.

e   Aileron and rudder controls great travel and small force increase in neutral position (typical English: Wellington and Lancaster).

f   Roll rate at 450 kph, one roll approximately 5 to 6 sec. Other data: takeoff weight 5,150 kg, not 6 tonnes as generally claimed. Wing area approximately 27.5 to 28 m², wing loading approximately 185 kg/m². Fuel capacity 729 l. Fuel type 100 octane, 130 octane for the Sabre VI. Armament four 20-mm Hispano 404 Mk. V cannon. Weapons arrangement: buried in wings. Wingspan 12.46 m. Overall length 10.24 m.

The Tempest shows a family resemblance to the Typhoon but structurally is very different and more closely resembles the Spitfire with its elliptical wing plan-

Hawker Tempest Mk. V. The Ta 152 was capable of engaging this British fighter successfully. The Tempest was tested in Rechlin towards the end of the war.

form. It is noteworthy that it has the sliding canopy of the 190, however since the pilot sits higher visibility from the cockpit is somewhat better than in the 190. The sliding hood has the same head armor, but it is only 10 mm thick. The wing has a laminar profile. The vertical tail has been enlarged through the addition of a prominent fuselage fillet, which is similar to that of the B-17.

The mainwheel arrangement has been changed so that a tire change now only requires the tire with hub to be removed instead of the entire wheel. This makes quick tire changes possible.

The Tempest has allegedly been used against the V 1 device, and 600 V 1s are said to have been brought down through the expedient of the Tempest overtaking the V 1 rocket in a dive and causing it to crash by making contact with its wing. An elastic device is probably fitted to the wing in order to avoid damage.

Rechlin, 16/3/1945 RB Schl/Ms

(Schlauch)

In Service with III/JG 301 and *Stab* JG 301

In the early days of December 1944 III *Gruppe* of *Jagdgeschwader* 301 learned that it was to convert to the new Focke-Wulf Ta 152 H. But as it turned out, III *Gruppe* was not to receive its first Ta 152 H until 27 January 1945.

Consequently the *Gruppe* continued to fly missions in the Defense of the Reich. While "Operation *Bodenplatte*" was taking place on 1 January 1945, JG 301 together with JG 300 formed the backbone of the Defense of the Reich. Beginning on 4 January 1945 III *Gruppe* moved to Alteno near Luckau and soon afterwards to Schroda near Posen, near the Russian front. The airfield had to be abandoned almost immediately as the front was drawing near. Most of the unit's Fw 190 A-8/R11s and Fw 190 A-8/R2s succeeded in taking off in spite of fog; the rest had to be blown up. The *Gruppe* subsequently returned to Alteno.

The day finally came on 27 January 1945. After the unit had relinquished its remaining Fw 190 A-8s to other units, it set off in wood-gas-powered trucks for Cottbus. In Neuhausen the pilots of III/JG 301 took charge of the first eleven Ta 152 H-0s. Their W*erknummern* were 150 001, 150 022, 150 025, 150 032 and 150 034-150 040. Following a short verbal indoctrination by Focke-Wulf technicians, the pilots flew the Ta 152 Hs back to Alteno. It was there that the only known photo of the unit's newly acquired Ta 152 H-0 high-altitude fighters was taken. But enemy action had already resulted in the destruction of fourteen brand-new Ta 152 H-0s at Neuhausen near Cottbus on 16 January 1945 and damage to two others[1] and these machines could not be delivered to the *Gruppe*. Even later, the original plan to equip III/JG 301 with 35 Ta 152 Hs was never brought to fruition.

Lineup of the first Ta 152 Hs of III/JG 301 at Alteno near Luckau on 27 January 1945 following the ferry flight from Neuhausen.

Captured Ta 152 H-1, *Werknummer* 150 168, "Green 9", already wearing British roundels as Air Min 11.

During its conversion until the end of February the *Gruppe* was not available for duty in the Defense of the Reich. At that point in time the *Gruppenkommandeur* was *Major* Guth, the adjutant Lt. Schröder and the technical officer Hptm. Hölzer (on 14 February 1945 Hptm. Hölzer was transferred to I *Gruppe*; his successor, Oblt. Schallenberg, came from II *Gruppe*).

During this conversion, on 1 February aircraft 150 037 was lost in a crash while being flown by Uffz. Hermann Dürr. Uffz. Dürr, of 12/JG 301, was on a practice mission not far from the airfield when the Ta 152 went into a flat spin. Uffz. Dürr was killed in the subsequent crash, which destroyed 98% of the aircraft. A second machine, 150 022, was made flyable again after a crash landing. By 14 February 1945 the *Gruppe* had already completed 120 flights totaling about 40 flying hours. All of III/JG 301's pilots were retrained on the new Ta 152 H and flew practice missions in the aircraft. At that time the *Gruppe* consisted of four *Staffeln* and as a rule each *Staffel* had twelve pilots. What was lacking was new Ta 152s. Aircraft serviceability averaged 75%, however this rapidly fell to 30% because water-contaminated fuel resulted

in injection pump seizures. The technical personnel soon had the problem in hand, however. Since the approaching front turned the Alteno airfield into an operational base for fighter and close-support *Gruppen*, III/JG 301 was supposed to move to Alperstedt near Erfurt so as to allow testing to go on uninterrupted. This did not come about, however, for on 16 February 1945 the *Gruppe* moved from Alteno to Sachau.

III/JG 301's first live action against American bomber units came on 2 March. The target of the Americans was the Böhlen chemical plant near Leuna. Twelve Ta 152s took part in this mission. The fighters' assembly point was the airspace between Burg and the Harz at an altitude of 8,000 meters. There was no contact with the enemy, however, since the Ta 152 pilots were forced to fend off repeated attacks by the Bf 109s of another unit. Fortunately there were no losses, as the climbing ability and maneuverability of the Ta 152s enabled them to evade these attacks. The unfamiliar shape of the Ta 152 was virtually unknown in the other *Jagdgeschwader*. A direct warning to the attacking fighters was not possible, for there was no direct radio contact between fighters. Each fighter unit was

[1] Confirmed by war diary entry, chief air armaments, air equipment working group.

Generalmajor Peltz, commander of the entire Defense of the Reich, inspects JG 301 on 14/03/45. From left to right: Obstlt. Fritz Auffhammer *Kommodore* JG 301, *Generalmajor* Peltz and *Gruppenkommandeur* Hptm. Herbert Nölter.

Another photo of Obstlt. Auffhammer, *Generalmajor* Peltz and Hptm. Nölter at Stendal; in the background is "Black 3", *Werknummer* 150 007 (the number "7" is just visible on the rudder), the aircraft of Obfw. Willi Reschke.

assigned its own frequency; this enabled radio communications with other aircraft in the formation but not other units. Communication with other units was only possible through the fighter control center, which in this case reacted too late. Among the pilots who took part in this failed mission were Hptm. Stahl, Lt. Reiche, Obfw. Sattler, Obfw. Keil, Fw. Reschke and Uffz. Blum.

A second mission a few days later also went badly. En route *Major* Guth's Ta 152 developed engine problems. Obfw. Reschke and Fw. Blum were ordered to escort *Major* Guth back to base. Since it was becoming obvious that the *Gruppe* was probably not going to reach its authorized strength of thirty-five Ta 152 Hs and that the majority of III *Gruppe*'s pilots could not be committed due to lack of aircraft, orders were issued to hand over all remaining Ta 152s to the *Geschwaderstab*, which at that time was based with II *Gruppe* at Stendal. One further Ta 152 was lost during this re-

organization. The machine of Obfhr. Jonny Wiegeshoff stalled and crashed while on approach, killing the pilot. Obfw. Reschke described the crash of Obfhr. Wiegeshoff: "The aircraft was visibly slower as it approached the airfield. Although his airspeed was obviously too low, he pulled up again at the airfield boundary and then came down like a stone. It is very likely that the propeller pitch control was no longer working and the propeller was in the feathered position." Then on 13 March 1945 came a strongly-worded order to hand over the remaining Ta 152 Hs to the *Geschwaderstab*. And so at 4:10 PM on 13 March 1945 Obfw. Willi Reschke took off from Sachau, landing at Stendal at 4:25. The next day III/JG 301 was visited by a high-ranking delegation under *Generalmajor* Peltz. At that point the entire Defense of the Reich was under Peltz's command and he even took the opportunity to fly Obfw. Reschke's Ta 152 H himself. Meanwhile III *Gruppe* was re-

equipped with new Fw 190 A-9 fighters and returned to action. On 14 March 1945 the *Gruppenkommandeur* of III/JG 301 *Major* Guth took over a *Luftwaffe* field battalion at Hagenow. His successor was *Hauptmann* Gerhard Posselmann.

From this time on the *Stabsschwarm* flew combined missions with II *Gruppe* from Stendal and provided top cover for II *Gruppe* during takeoff and landing. In coordination with the base flak, pairs of Ta 152s took off to protect the *Gruppe* from attacks by enemy fighters and fighter-bombers. The *Stabsschwarm* consisted of the following pilots: Obstlt. Auffhammer, Hptm. Stahl, Obfw. Sattler, Obfw. Keil, Obfw. Reschke and Fw. Blum. After the transfer to Neustadt-Glewe they were joined by Obfw. Loos. During this period II *Gruppe* lost just one aircraft, a Fw 190 D-9 shot down by a P-47 Thunderbolt while attempting to land.

This was a very difficult time for the handful of Ta 152 pilots, for dealing with at-tacks from all sides while constantly out-numbered was sometimes impossible. But the Ta 152 demonstrated its qualities as a fighter in these defensive battles. Its high speed, tight turning radius and enormous climb rate must actually have brought many P-47 and Tempest pilots to the point of des-peration. Not a single Ta 152 H was lost in all these airfield defense missions. On 7 April 1945 the *Stabsschwarm* received word that two Ta 152s were sitting at the Rewe 2 production site at Erfurt-North waiting to be picked up. Early on the morn-ing of 8 April Obfw. Reschke and Fw. Blum took off in an Ar 96. The air raid sirens were howling as the two brand-new Ta 152s left Erfurt for the flight back to Stendal. Obfw. Reschke described this daring mission: "Both machines were unarmed and it wasn't so easy to get them to Stendal in one piece." On 10 April Erfurt was taken by the Americans. Unfortunately it is no longer possible to ascertain the variant and

Generalmajor Peltz climbs into the cockpit of Ta 152 H "Black 3". Clearly visible are the octane triangles for the fore and aft tanks and the border for the 115-l auxiliary tank or GM 1 tank.

Another photo of *Generalmajor* Peltz in the cockpit of the Ta 152.

Hptm. Nölter reads out the order of the day at Stendal on 15 March 1945; in the background Fw 190 D-9s of II/JG 301 and on the far right a Ta 152 H.

Werknummern of these Erfurt Ta 152s. The Ar 96 was left behind in Erfurt. The *Stabsschwarm* was now back up to eight serviceable Ta 152s. That same day the Ta 152 Hs engaged fifteen P-47s in the Brunswick area, in the course of which Jupp Keil was able to shoot down a Thunderbolt. Since operational conditions were changing from day to day, JG 301 and its three *Gruppen* moved from Stendal, Salzwedel and Sachau to the airfields at Neustadt-Glewe, Ludwigslust and Hagenow. The *Stabsschwarm* flew its last missions with the Ta 152 H from Neustadt-Glewe. Obfw. Reschke: "Neustadt-Glewe was a fortunate choice, since the airfield was surrounded by light, twin-barreled anti-aircraft guns which gave our machines the necessary protection during takeoff and landing, preventing the ‚Hackers', as the enemy fighter-bombers were called, from attacking the field at low level." The *Stabsschwarm* suffered its first loss at Neustadt-Glewe. During the afternoon of 15 April 1945 several British Tempests strafed rail installations in Ludwigslust. Four Ta 152s took off to intercept. The Ta 152 H of Obfw. Sepp Sattler

was lost before combat was joined for reasons unknown. The remaining Ta 152s engaged the Tempests of No. 486 Squadron at low level. Obfw. Willi Reschke positioned his "White 1" behind the Tempest being flown by Lt. Mitchell and damaged its tail assembly with his first burst. Thus warned, the Tempest pilot, now at ground level, tried to escape the Ta 152 H's field of fire by turning ever tighter. While this posed no problem for the Ta 152 H, the Tempest stalled and crashed in a nearby wooded area. Obstlt. Fritz Auffhammer was very fortunate after his Jumo 213 E suffered supercharger failure and visibly lost power during combat. Bathed in sweat, he managed to land at Neustadt-Glewe. Pilots Obfw. Sattler and Lt. Mitchell were buried side by side in Neustadt-Glewe.

On 24 April 1945, during the Battle of Berlin, there were engagements with Yak 9s. Once again the *Stabsschwarm* put up three pairs of fighters. These accompanied II *Gruppe*, which carried out low-level attacks against Russian positions. Since there was no contact with enemy aircraft, the *Stabsschwarm* was ordered to reconnoiter

Obfw. Reschke scored
several victories while
flying the Ta 152 H.

Oberfeldwebel Josef
Keil scored five victo-
ries with the Ta 152 H
and shot down the first
Mustang on 1 March
1945.

over Berlin. There was an encounter with Yak 9s in poor visibility. In the ensuing dog-fight the *Stabsschwarm* lost Hptm. Hermann Stahl and his Ta 152. Four Yak 9s were shot down, two by Obfw. Willi Reschke and two by Obfw. Walter Loos (Green 4). The disarray in command of the last days of the war makes it impossible to date the last mission with certainty, but it was probably on 30 April 1945. On that day Walter Loos shot down another Yak 9. The Ta 152s had shot down at least ten enemy fighters for the loss of just two of their own number. The last transfer of the *Stabsschwarm* was to Leck in Schleswig-Holstein. All remaining Ta 152s were handed over to the British. While other Ta 152s were scrapped, "Green 9" (WNr. 150 168) was spared and was flown to England in the belly of an Ar 232.

This Ta 152 was last flown by Obfw. Willi Reschke. The Ta 152 H-1, now designated Air Min 11, was tested by famous test pilot Eric Brown during a flight from Farnborough to Brize Norton. Curiously, the British experts did not succeed in evaluating the Ta 152 H-1's performance with GM

1 and MW 50 injection. They were satisfied with what they had, consequently there were no performance comparisons with Allied fighter aircraft. Ta 152 *Werknummer* 150 168 was scrapped in 1946. Unfortunately there is no information as to the fate of other Ta 152s captured by the British. The chapter on the operational career of the Ta 152 ends with a sentence by Willi Reschke. "The Ta 152 was my life insurance in the last days of the war." This sentence reflects the enormous relationship of trust that existed between the Ta 152 pilots and their machines.

Remarks

III *Gruppe* used the following *Staffel* markings:

9 *Staffel*:	white numbers
10 *Staffel*:	red numbers
11 *Staffel*:	yellow numbers
12 *Staffel*:	black numbers

Aircraft of the *Stabsschwarm* wore green numbers. In JG 301 it was not standard practice to apply the *Werknummer* as well as the tactical number (eg. Green 9), thus it is impossible to associate a pilot with

Oberfeldwebel Gerhard Loos scored several victories with the Ta 152 H.

a particular aircraft. Consequently, the names of all the pilots known to have flown the Ta 152 on operations are listed here. I would be grateful for any additional information.

Operational Ta 152 H Pilots of JG 301

Major Guth
Obstlt. Fritz Auffhammer
Oblt. Schallenberg
Lt. Reiche
Hptm. Hermann Stahl
(killed in air combat 24/4/1945)
Obfw. Sepp Sattler
(killed in action 15/4/1945)
Obfw. Josef Keil
Obfw. Walter Loos
(38 victories, including 22 heavy bombers and 8 Soviet aircraft)
Obfw. Willi Reschke
(26 victories, including 18 heavy bombers)
Obfw. Herbert Stephan
Fw. Bubi Blum
Obfhr. Jonny Wiegeshoff
(killed in the crash of his Ta 152 on 14/3/1945)
Uffz. Hermann Dürr
(killed in the crash of his Ta 152 on 1/2/1945)

Jagdstaffel 152 and Operations with *Stab* JG 11

The first Ta 152 H-0s from the Cottbus production line were handed over to the Rechlin Test Station for trials. It was intended that the first production aircraft would be used for trials at Rechlin since full-scale testing with the four non-production prototypes (the Fw 190 V 33/U1, V/30 U1, V 29/U1 and V 18/U1) had not been possible. Consequently it was anticipated that there would be delays resulting from complaints about flight safety and handling characteristics as well as a considerable number of initial changes. The Rechlin Test Station requested a total of twelve Ta 152 H-0s for trials. The first aircraft arrived at Rechlin on 11 December 1944. All test aircraft were delivered by 31 December (22 December—1 Ta 152 H, 23 December—1 Ta 152 H, 29 December—1 Ta 152 H, 30 December—4 Ta 152 H, 31 December—4 Ta 152 H). In command of *Erprobungskommando* Ta 152, EK 152 for short, was *Hauptmann* Bruno Stolle. Stolle had arrived from I/JG 11 on 25 November 1944 and had a total of 35 victories to his credit, including 5 heavy bombers. Later, EK 152 was officially redesignated *Stabsstaffel* JG 301, but it never joined III/JG 301. Trials were supposed to last until 1 April 1945. Original plans to expand EK 152 into four *Staffeln* never came about. Instead III *Gruppe* of JG 301 was involved in the testing process.

Later, the desperate military situation led to the order that all of the test station's aircraft were to be used against the enemy. The so-called *Gefechtsverband K.d.E.* (Test Command Battle Unit) was supposed to be formed at Rechlin and was under the command of *Oberst* Petersen, director of the Rechlin Test Station. All operational Me 262s, Ar 234s, Bf 109s, Fw 190s, Ta 152s, Ju 88s, Ju 188s and He 111s and their crews were to be part of the unit. According to a strength report dated 9 February 1945 the *Gefechtsverband K.d.E.* had on strength 9 Bf 109s, 4 Me 262s, 8 Ta 152s, 25 Fw 190s (two *Jabostaffeln*), 9 Ar 234s (high-speed bomber *Staffel*), 10 He 111s and 17 Ju 88s and Ju 188s.

Jagdstaffel Ta 152, formerly EK 152, was to fly operations from Roggentin. *Hauptmann* Bruno Stolle remained in command of the *Staffel*. On 4 February 1945 the *Staffel*, now designated "*Jagdstaffel Roggentin*", was made available by the K.d.E. for operations in the east. At this point the number of aircraft was six Ta 152 H, which rose to eight on 8 February. So far it is not known whether this unit was used operationally, however the *Werknummern* of seven Ta 152s at Rechlin are known. The aircraft were pre-production Ta 152 H with the *Werknummern* 150 003 (CW + CC), 150 006 (CW + CF), 150 008, 150 009 (CW + CI), 150 010 (CW + CJ), 150 011 and the Ta 152 C V8, WNr. 110 008 (GW + QA). It is also known that four to six Ta 152 H were assigned to the *Stab* of JG 11 at Neustadt-Glewe at the end of April 1945 shortly before its move to Leck. According to information provided by *Herr* Mehling, then a *Leutnant* in *Stab* JG 11, all appeared to have been test aircraft since no two were alike.

Front view of the captured Ta 152 H-0 *Werknummer* 150 010.

Hptm. Bruno Stolle commanded the test detachment EK 152, which was equipped with the Ta 152 H, at Rechlin.

The Ta 152s were not used in combat and only a few familiarization flights were carried out. During its last transfer to Leck the unit was engaged by Spitfires, resulting in the loss of two Ta 152 H. A third made a belly landing at Leck, as a result of which just

Side view of the tenth Ta 152 H-0, 150 010.

150 010 with engine running.

one Ta 152 H, flown by Lt. Mehling, landed safely there. Thus ended the wartime career of the Ta 152 H.

Plans for a comparison flight on 9-10 May 1945 involving the Ta 152 H flown by Lt. Mehling and a Spitfire XIX were canceled for security reasons. It was feared that the German pilot might parachute from the Ta 152, resulting in the loss of the aircraft. Lt. Mehling recalled that, compared to the

Fw 190 D-9 he had previously flown, the Ta 152 H could climb better, was more maneuverable and had a lower landing speed.

The Ta 152s with the W*erknummern* 150 003 (CW + CC) and 150 010 (CW + CJ) supposedly fell into British hands intact. While no information can be found as to the fate of 150 003, after the war 150 010 was taken to America, where it remains as the sole existing Ta 152 H-0. 150 010 was cap-

tured by the British at Aalborg (Denmark) and was later passed on to the Americans. It initially received the marking "USA 11" and was ferried to Melun, France by pilot Fred McIntosh of the "Watson Whizzers." After crossing the Atlantic on the carrier *Reaper*, the Ta 152 reached Wright Air Force Base in Indiana, where it was designated FE 112 (FE=Foreign Evaluation). This was later changed to T2-112. The code "Green 4" applied in America, which was probably supposed to indicate the *Stabsschwarm* of JG 301, cannot be verified. Photographs reveal that in addition to its factory code of CW + CJ the aircraft probably wore a "2" at some point as well as a black "6" or "8." Today the disassembled Ta 152 is in the possession of the NASM in Washington and there are no plans to restore it in the foreseeable future. Although the Americans captured at least two Ta 152s (including 150 167) and apparently made them ready to transport, there is no information as to their fates, even from the American side.

Production of the Ta 152 H at Cottbus

Production of the Ta 152 H by Focke-Wulf began with the Ta 152 H-0 version, which had no wing tanks and no power boosting (MW 50 or GM 1). Assembly of the Ta 152 H-0 began in the Cottbus factory. The first two production Ta 152 H were test flown at the end of November 1944. Test pilot Hans Sander flew the first production Ta 152 H-0 (WNr. 150 001, CW + CA) on 29 November 1944. Sander then test flew the second production Ta 152 H-0 (WNr. 150 002, CW + CB) on 29 November and the third (WNr. 150 003, CW + CC) on 3 December. Eighteen more Ta 152 H-0s were accepted in December 1944. During this time *Werknummer* 150 025 made a crash landing on Cottbus airfield. Material failure caused the undercarriage bolts to shear off. This machine was repaired, however, and was one of those handed over to III/JG 301 on 27 January 1945. Twenty more Ta 152 H were accepted in January 1945 and the last three in February. After this production at Cottbus ended for good due to missing components and relocation measures. It is thus very likely that a total of just forty-three Ta 152 H were produced in Cottbus. On 16 January 1945 Neuhausen was attacked by approximately forty Lightnings and Mustangs, destroying fourteen brand-new Ta 152 H and damaging another. This loss had serious consequences for III/JG 301, preventing it from reaching its planned strength of thirty-five Ta 152 H fighters.

One Ta 152 H (Fw 190 V 18/U2, CF + OY) was used for training purposes, especially for the factory test pilots, from 21 to 25 November 1944. Then on 25 November Hans Sander flew the V 18/U2 back to Langenhagen. After this, from 28 November to

Lt. Mehling flew his last mission in a Ta 152 H from Neustadt-Glewe to Leck. Here he is seen with air movements clerk Kurt Brauer on the tail of his Fw 190 A "White 4" while with 7/JG 51.

Lt. Mehling leaves the cockpit of his Fw 190 A after a combat mission at the end of 1943. He ended the war with 22 confirmed victories, including 2 (3) Thunderbolts which he shot down on 4 January 1945 while flying a Fw 190 D-9.

3 December 1944, the Fw 190 V 29/U1, GH + KS, was used for further training at Cottbus. On 23 December the V 18/U2 ground-looped to the right while taking off, as a result of which the right undercarriage leg retraction cylinder attachment fitting was torn off. The new wooden tail was first fitted during the repair process.

The Low-Level Attack on Neuhausen Airfield on 16 January 1945

On 16 January 1945 the Ta 152 H production program suffered a severe setback. The following account is from a report addressed to the OKL, Chief of Air Armaments.

Time of the attack: 12:03 to 12:35 PM.

Number and type of enemy aircraft: several Mustangs and Lightnings in the Ruhland-Neuhausen area.

Height and execution of attack:

Aircraft flew over the airfield at a height of about 500 m and flew away to the east without firing. Then followed a low-level attack from the northeast and southwest with subsequent passes from south to north and west to east, sometimes as low as two to three meters above the ground. The attack was made in approximately four waves of about ten aircraft each against aircraft camouflaged on the airfield boundary and parked in the woods.

Weather: clear.

Makeshift defense on the airfield by *Luftwaffe* unit stationed there. Used were one 20-mm gun, ten MG 81 machine-guns.

Results of attack: Focke-Wulf company aircraft destroyed: 14 Ta 152, 1 Fw 190. One Ta 152 approximately 30% damage.

Facilities destroyed: One small wooden hangar with accumulator station housed there, transformers and work benches. (Destroyed by fire.)

The aircraft cited as destroyed in the report were parked in the forest or close to the forest edge around the airfield. Camouflage was enhanced through the use of fir branches and in some cases camouflage nets. The blast pens in the Neuhausen dispersal area could only be used by the Fw 190 and not by the Ta 152 on account of the latter type's wingspan.

Ta 152 H-0 *Werknummer* **150 005, CW + CE, on the compass-swing platform at Cottbus.**

Front view of the fifth production aircraft, 150 005, which was test flown in December and later served Junkers in the Jumo 213 E test program.

The large number of completed aircraft present at Neuhausen can be explained by the fact that two special tests have to be carried out following completion of the aircraft and testing by the B.A.L.:

1. Testing of the motor attachment bolts in Fw 190 Replacement Part No. 190.641-0106. 2. Check for cracks in weld seams on aileron pushrods. Carrying out these checks and replacing defective parts takes some time, however for reasons of flight safety the aircraft cannot be ferried away before the checks are done. Finally another check has to be carried out on the

Planned Production of Individual Variants of the Ta 152 from 11 January 1945

Baureihe	Verw.-Zweck	Motor	Firma	1945							
				1	2	3	4	5	6	7	8
C-1/R11	J	DB 603 L	MMW	–	10	20	30	30	40	50	–
			ATG	–	–	5	25	50	120	120	–
			SFH	–	–	5	15	50	–		
			GFW	–	–	–	–	30	30	–	
			WFG	–	–	–	–	10	30	80	150
B-5/R11	Z	8213 F-1	Erla	–	–	–	–	5	30	80	120
E-2/R11	A	8213 F-1	MMW	–	–	10	20	50	70	80	–
H-1/R11	J Hö	8213 F-1	FW	10	30	100	200	350	–	–	–
			Erla	–	–	3	12	35	100	–	–
			WFG	–	–	2	8	15	30	50	–

Firmen:
ATG   Allgemeine Transportanlagen Gesellschaft in Leipzig;
MMW   Mitteldeutsche Metallwerke in Erfurt; Erla in Leipzig
GFW   Gerhard Fieseler Werke in Kassel;
SFH   Siebel Flugzeugbau in Halle;
WFG   Waggonfabrik Gotha

Werknummer List of Production Aircraft Built at Cottbus

Werknummer	Registration	Remarks
150 001	CW + CA	First flight on 24 November 1944 at Cottbus, pilot Hans Sander. Forced landing with engine stopped. Used as operational aircraft by III/JG 301 from 27/1/1945.
150 002	CW + C B	First flight on 29 November 1944 at Cottbus, pilot Hans Sander
150 003	CW + CC	First flight on 3 December 1944 at Cottbus, pilot Hans Sander. First Ta 152 H to Rechlin on 11 December 1944. Converted to wooden tail at Rechlin, from 4 February 1945 with Jagdstaffel Ta 152. Staf*felkapitän* Hptm. Stolle, based at Roggentin
150 004	CW + CD	First flight on 17 December 1944 at Langenhagen, pilot Hans Sander. Test-bed at Langenhagen, last known flight on 19 February 1945. Converted to strengthened cowling for increased speed and test flown by pilot Schnier at Langenhagen on 9 February 1945. Aircraft set aside at Reinsehlen on 9 February 1945
150 005	CW + C E	Circuit at Cottbus on 8 December 1944, assigned to Junkers, Dessau for engine trials, still there on 18 March 1945
150 006	CW + CF	Test flight from Neuhausen to Cottbus on 27 December 1944, Cottbus to Neuhausen on 31 December 1944. Picked up at Neuhausen on 31 December. Known to have been tested at Rechlin from 10 February to 2 march 1945. Operational aircraft with Jagdstaffel Ta 152.
150 007	CW + CG	"Black 3," first III/JG 301 later Sta*bsschwarm* JG 301, test flown by GMJ. Peltz on 15 March 1945. Pilot Obfw. Reschke
150 008	CW + CH	Testing at Rechlin. Belly landing at Kleinhausen on 20 February 1945, pilot Baist. Operational aircraft Jagdstaffel Ta 152.
150 009	CW + CI	Factory test flight Cottbus on 17 December 1944. 24 December 1944 ferried from Cottbus to Roggentin by pilot Kamp. Operational aircraft Jagdstaffel Ta 152, then to Stab JG 11.
150 010	CW + CJ	Tested at Rechlin from 30 January to 8 March 1945. Second Ta 152 H with wooden tail. Operational aircraft Jagdstaffel Ta 152, then to Stab JG 11
150 011	CW + CK	Testing at Rechlin, first to have GM 1 installed. Operational aircraft Jagdstaffel Ta 152
150 012		No information
150 013		Check flight Cottbus to Neuhausen on 2 January 1945
150 014		First flight on 23 December 1944, pilot Bielefeld (factory pilot). Test flight from Neuhausen to Cottbus on 29 December 1944. Acceptance flight Cottbus to Neuhausen on 5 January 1945
150 015		Check flight Neuhausen to Cottbus on 5 January 1945. Check flight at Neuhausen on 6 January.
150 016		Acceptance flight from Neuhausen to Cottbus on 29 December 1944. Check flight from Cottbus to Neuhausen on 3 January 1945
150 017		Test flight from Cottbus to Neuhausen on 29 December 1944. Check flight from Neuhausen to Cottbus on 3 January 1945.
150 018		No information
150 019		First flight on 29 December 1944 at Neuhausen, test pilot Bielefeld
150 020	CW + CT	Test flight Neuhausen on 10 January 1945
150 021		Check flight Neuhausen to Cottbus on 31 December 1944. Check flight Cottbus to Neuhausen on 4 January 1945.
150 022		Accepted Neuhausen on 10 January 1945. From 27 January operational aircraft with III/JG 301. Crash landing in February 1945, repaired.
150 023		First flight from Cottbus to Neuhausen on 29 December 1944, pilot Bielefeld. Crashed while being flown by Hptm. Eggers on 9 February 1945 during ferry flight from Tarnewitz to Rechlin.
150 024		First flight from Cottbus to Neuhausen on 31 December 1944, pilot Bielefeld.
150 025		Check flight Neuhausen to Cottbus on 31 December 1944, undercarriage failure on landing (10%). Check flight Neuhausen on 28 January 1945. From 27-28/1/1945 operational aircraft with III/JG 301
150 026		No information
150 027		Test flight Neuhausen on 5 January 1945. Converted as test-bed for Ta 152 C-3 with MK 103 engine-mounted cannon, DB 603 E engine.
150 028		No information
150 029		First flight on 7 January 1945 at Cottbus.
150 030		Flights on 1 and 2 February 1945 at Langenhagen, pilot Hans Sander. Converted as weapons test-bed for Ta 152 C-3 with MK 103, DB 603 E engine.
150 031		No information

Werknummer	Registration	Remarks
150 032		First flight on 17 January 1945 at Cottbus, from 27 January operational aircraft with III/JG 301.
150 033		No information.
150 034		First flight Cottbus to Neuhausen on 20 January 1945. Test flight at Neuhausen on 23 January. Accepted at Neuhausen on 23 January. From 27 January operational aircraft with III/JG 301.
150 035		From 27 January 1945 operational aircraft with III/JG 301.
150 036		First flight at Cottbus on 16 January 1945, from 27 January operational aircraft with III/JG 301.
150 037		First flight Cottbus to Neuhausen on 18 January 1945, from 27 January operational aircraft with III/JG 301. Crashed on 1 February 1945, pilot Uffz. Dürr killed, 98% write-off.
150 038		From 27 January 1945 operational aircraft with III/JG 301.
150 039		From 27 January 1945 operational aircraft with III/JG 301.
150 040		From 27 January 1945 operational aircraft with III/JG 301.
150 167		Captured in flyable condition at Erfurt-North by American troops on 15 April 1945; in all likelihood 150 167 was earmarked for conversion to Ta 152 H-10 standard.
150 168		"Green 9." Last Luftwaffe flight by Obfw. Willi Reschke. Captured at Leck and taken to England, there flown from Farnborough (test station) to Brize Norton by Eric Brown, scrapped in England.
150 169		Highest known Werknummer, supposedly also captured at Leck.

Front view of *Werk-nummer* 150 010 at Wright Field in America as captured enemy aircraft FE 112.

Rear view of CW + CJ with FE (foreign evaluation) number 112 on its tail while under evaluation in the USA.

A Ta 152 minus propeller, parked at Leck; this captured Ta 152 H was supposedly *Werk-nummer* 150 169.

fuel tanks, as received by telex from the officer at the Rechlin Test Station in charge of the type, Hptm. Schmitz.

It should also be mentioned here that a wider dispersal of the aircraft was not possible because the airfield was also being used by a *Luftwaffe* unit (SG 151).

As stated at the beginning, the enemy did not fire on his first pass. The attack then followed at low-level from various directions uninterrupted against every part of the airfield; the parked aircraft were strafed individually.

According to an eyewitness report, the enemy fighters' attention was attracted by two training aircraft landing at Neuhausen. These two training aircraft set about landing at Neuhausen during the alert period preceding the attack.

Cottbus, 18 January 1945

The Collapse of Production

The outlook for the future was not good when production of the Focke-Wulf Ta 152 H began in November 1944. The growing Allied air attacks from the west had led Focke-Wulf to decentralize production of all types as much as possible and transfer production as far east as possible beyond the range of Allied bombers. Now, with the Eastern Front drawing nearer, these production sites were seriously threatened. The original roll-out date for the first Ta 152 H-0, the beginning of November 1944, was delayed by difficulties in production start up caused by inaccurate blueprints and missing jigs. The jigs ordered in France had been lost in the summer of 1944. Plans to produce the Ta 152 in Italy, reached on 9 May 1944, were dropped on 24 July of the

Two dismantled Ta 152s and a Fw 190 D-9; in the background is *Werknummer* 150 167, in the foreground a Fw 190 D-9 of JG 301.

same year. Sealing the pressurized cockpit caused great problems in production of the high-altitude fighter. Thus it was 11 December 1944 before the first Ta 152 H-0, *Werknummer* 150 003, reached the Rechlin Test Station.

In order to speed up production of the Ta 152 H, on 11 January 1945 the armaments staff decided to concentrate efforts on the Me 262 (8-262) under the central control of Director Thiedemann and the Ta 152 (8-152) under the central control of Dr. Reichelt.[1] At the same time the quality of the product, including the Ta 152, continued to decline. Poor factory work even led to a halt in production for a short time. The cause was poor welding on the aileron pushrods. In January 1945 the Aircraft Testing Group of the Director of Air Armaments decided that henceforth the Daimler Benz DB 603 engine would only be used for the Fw 190 series, with the intention of outfitting fifteen Fw 190 D-11s with the DB 603 in January 1945 and fifteen Fw 190 D-12s in February. The Arado firm was to deliver the entire Fw 190 D-14 series.

The first breakdowns in production occurred at the end of January 1945. There were no deliveries of wings and hydraulic systems. Fuselage and wing production in Posen were lost and with it the jigs and tools. The planned testing of the Ta 152 C series with the more powerful DB 603 LA engine was so badly delayed that at the end of January it was assumed that production could not begin before March-April 1945. The reason for this was urgently needed engine modifications.

A new emergency program was adopted at the *Reichsmarschall*"s conference of

[1] Dr. Reichelt was director of the Arado Werke.

Side view of 150 168, ex "Green 9", during a display of captured aircraft in England in 1946.

Rear view of Ta 152 H-1 150 168; in England this aircraft was evaluated briefly by the well-known test pilot Eric Brown.

22 February 1945. Ta 152 variants were severely limited with the proviso that all Ta 152s were to go to fighter units, with the close-support units bolstered by the addition of converted Fw 190s. Then at the beginning of March 1945 there were unexpected problems with the Ta 152's longitudinal stability.

To address this the Aircraft Testing Group ordered the planned 140-liter MW 50 tank limited to 75 liters. A further improvement was to be achieved by increasing the size of the tail surfaces and modifying the wing-fuselage junction. In mid-March another serious problem was encountered with the MW 50 tanks for instal-

lation in the wing. Focke-Wulf's plans to switch the GM 1 system in the fuselage to MW 50 did not meet with the approval of the OKL.

Then, at a meting of the armaments staff on 29 March 1945, it was finally proposed that the Ta 152 program be set aside in order to allow the Fw 190 D with the Jumo 213 F engine to proceed. The reason for this was the total collapse of the start-up of series production and the lost production areas. No resumption of production was planned, in order not to threaten production capacity for the Me 262.

Far in excess of 15,000 Ta 152s were to have been produced by March 1946, however only a handful were in fact completed. Curiously, in April 1945 all of the blueprints for the Focke-Wulf Ta 152 were sold to Japan; however, no production of the Ta 152 was undertaken there.

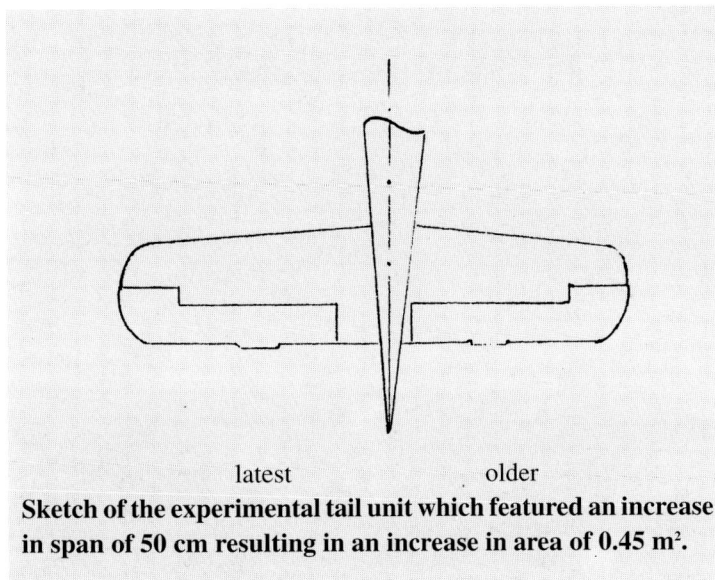

latest   older

Sketch of the experimental tail unit which featured an increase in span of 50 cm resulting in an increase in area of 0.45 m².

Development of the Focke-Wulf Ta 152 Reconnaissance Aircraft

In addition to the standard fighter version, the OKL[1] also requested standard and high-altitude reconnaissance variants of the Focke-Wulf Ta 152.[2] The standard reconnaissance version was designated the Ta 152 E-1 and was supposed to replace current fighter-reconnaissance versions of the Bf 109. The Ta 152 E-1 was based on the Ta 152 B (Jumo 213 E) or Ta 152 C (DB 603 L) (wing area 19.5m²).

Initially the following basic equipment was anticipated for the Ta 152 E-1:

a   Power plant Jumo 213 E

b   Armament MK 108 engine-mounted cannon and MG 151s in the wing roots.

c   Cameras RB 75/30, or RB 50/30, or RB 30/18, or RB 50/18 or two staggered RB 20/12x12, or two staggered RB 40/12x12, or two staggered RB 12.5/7x9, or two staggered RB 32/7x9. It was anticipated that a camera rack would be added. The framework was to be built by the units.

d   Robot II camera in wing leading edge.

e   FuG 15 and 25a

f   Type 67 G periscope

g   Fuel tanks in wings

h   Choice of MW 50 or extra fuel in fuselage (115 l tank)

i   Type 503 external stores rack under fuselage for carriage of 300-l drop tank (carriage of bombs not anticipated).

In the course of development the high-altitude reconnaissance version, which was originally designated the Ta 152 E-2, became the Ta 152 H-10 and was based on the Ta 152 H-1 high-altitude fighter. It shared the same basic configuration as the Ta 152 E-1 but had a pressurized cockpit, the larger wing (23.5m²) and, in the course of further development, the MW 50 system.

Another interesting proposed version was the Ta 152 E-1/R1. It was to carry a RB 50/18 camera mounted at an angle, specifically 10 degrees to the horizontal. In order to accommodate this a bulge was added to

Front view of Ta 152 H-0 150 010 in America, now as T2-112.

[1] Luftwaffe High Command
[2] Focke-Wulf development report Ta 152 E No. XVIII a1-a3 dated 24/1/1945

Side view of T2-112 with propeller removed.

Type Description No. 282 Ta 152 E dated 12 July 1944

General data pertaining to the Ta 152 E-1 and Ta 152 E-2 (Ta 152 H-10)

	Ta 152 E-1	Ta 152 E-2 (Ta 152 H-10)
Purpose:	Single-seat reconnaissance aircraft for medium altitudes without pressurized cockpit	Single-seat reconnaissance aircraft for high altitudes with pressurized cockpit
Configuration:	Single-engined, low-wing cantilever monoplane with hydraulically retractable undercarriage.	
Structural strength:	Maneuvering load factor 6.5 at a designed takeoff weight of 4,500 kg	Maneuvering load factor 5.0 or –2.5 at a designed takeoff weight of 4,400 kg
Power plant:	Jumo 213 E standard power plant	

Dimensions:			
	Wing area:	19.6 m²	23.5 m²
	Wingspan:	11 m	14.82 m
	Aspect ratio:	6.17	9.4
	Vertical tail area:	1.77 m²	1.77 m²
	Horizontal tail area:	2.82 m²	2.82 m²
	Maximum length:	10.810 m	10.810 m
	Maximum height:	3.360 m	3.360 m

Normal takeoff weight:	4 675 kg with Jumo 213 E and MW 50 system	4 815 kg with Jumo 213 E and GM 1 system

Armament: 2 MG 151/20 with 175 rounds per gun
1 MK 108 with 85 rounds

Cameras and optical aids: 1. In fuselage
RB 75/30
RB 50/30
1 x RB 30/18 or RB 50/18
2 x RB 20/12x12 staggered
2 x RB 40/12x12 staggered
2 x RB 32/7x9 staggered
2. In leading edge of left wing: 1 Robot II
3. In cockpit: periscope (Voigtländer)

121

Close-up view of the Ta 152 H with firewall access panels open. Clearly visible are engine attachment points and oil and hydraulic fluid reservoirs.

one side of the fuselage. The necessary camera housing was tested on a standard production Fw 190 D-9.

Ta 152 E-1 Reconnaissance Aircraft

A reconnaissance aircraft designated the Ta 152 E-1 was developed from the standard Ta 152 B fighter based on the guideline GL-C-E2 No. 11303/44 (IIIa) of 28/3/44. All of the stated requirements could be met satisfactorily with relatively minor modifications, which mainly affected the rear fuselage. The following is a description of the significant changes to the Ta 152 B airframe (also see Type Description Ta 152 A and B No. 270).

Important Design Changes to the Airframe of the Ta 152 B

The requirement for installation of the methanol-water tank or extra fuel tank while simultaneously taking into consideration seven different camera configurations made necessary the following changes to the fuselage:

a   Movement of Bulkheads 9 and 10 to the rear. The bulkheads were designed as attachment bulkheads for the camera mounts.

b   Movement and enlargement of the fuselage access hatch on the left side wall. Equipment installation and cassette removal and installation took place through this hatch. Dimensions were such that installation of the methanol-water tank was also possible. This made it possible to do away with the hatch on the underside of the fuselage. A small fuselage cutout was required for the camera window.

c   Installation of cameras made necessary the relocation of the following items of equipment: the FuG 25a IFF set had to be moved one bay to the rear, the DF loop was positioned between Bulkheads 8 and 9.

d   The tail-heavy moment created by the heavy camera installation was compensated for by relocating the bottles previously housed in the fuselage extension. Installed in the left wing root were: one 2-liter compressed air bottle for emergency lowering of the landing flaps and undercarriage and one 5-liter compressed air bottle for the MK 108;

in the right wing root: three oxygen bottles and one 2-liter compressed air bottle for emergency lowering of the landing flaps and undercarriage.

Cameras

The measures described above made it possible to install the following camera equipment in the fuselage; installation and changing of cassettes was easily accomplished through the large side hatch:
a) Rb 75/30 vertical installation
b) Rb 50/30 vertical installation
c) 1 x 30/18 or
    1 x 50/18 vertical installation
d) 2 x Rb 20/12x12 staggered
e) 2 x Rb 40/12x12 staggered
f) 2 x Rb 12.5/7x9 staggered
g) 2 x Rb 32/7x9 staggered

Periscope

A periscope made by the Voigtländer firm was installed in the cockpit to enable the pilot to observe the terrain directly beneath him. The periscope passed between the two fuel tanks and left of the control stick into the cockpit. As a result there was a minor modification of the rear tank, which was now in general use in the basic Ta 152. In order to make possible the pressurized cockpit of the Ta 152 E-2 a pressure-tight conduit through the floor was anticipated, consisting of pressure-tight bushings.

Robot II Camera

The Robot II was installed in the leading edge of the left wing in place of the BSK 16 gun camera, and an insert was planned to accommodate it. Electrics and control button remained the same.

Armament

Anticipated armament consisted of a MK 108 engine-mounted cannon with 85 rounds and two MG 151 in the wing roots with 175 rounds per gun. If necessary, two MG 151 with 150 rounds per gun could be installed in the fuselage in place of the engine-mounted cannon.

Radio Equipment

The reconnaissance version, production of which was supposed to begin in April 1945, was initially to be equipped with the FuG 16 ZS. Replacement by the FuG 15 Y was to take place as soon as the equipment became available. A FuG 25a IFF set was installed in the rear fuselage.

Oblique camera installation planned for the Ta 152 E-1/R1 series under test on Fw 190 D-9 210 002, TR + SB.

Auxiliary Tanks

The following auxiliary fuel tanks could be used to increase range:

1. In place of the MW 50 system of GM 1 tank a protected long-range tank holding 115 liters in the fuselage aft of the cockpit.
2. A 300-liter drop tank under the fuselage. The tank was mounted on the ETC 503 stores rack. The oil tank was of such dimensions that 25% cold start mixture was possible with the 115-liter auxiliary tank. Cold start mixture had to be dispensed with when the auxiliary fuel tank and the 300-liter drop tank was used together, while use of the 300-liter drop tank alone required a reduction in cold start mixture of about one half.

Performance Boosting

For increased performance a 140-liter methanol-water (MW 50) tank could be installed in the fuselage in place of the auxiliary fuel tank. The contents enabled an operating duration of approximately 42 minutes at takeoff and emergency power (average consumption 22 l/hr). The use of GM 1 was also possible, but would not have been an advantage at the low altitudes at which the Ta 152 E-1 standard reconnaissance aircraft was expected to operate.

Ta 152 E-2 High-altitude Reconnaissance Aircraft

Construction of a small number of Ta 152 E-2 high-altitude reconnaissance aircraft was planned for reconnaissance flights at extreme altitudes. In terms of cameras and armament it was identical to the standard Ta 152 E-1. The only differences were: the larger wing of the Ta 152 H high-altitude fighter (23.5 m²) was adopted for better high-altitude performance. As well the fuselage was equipped with a pressurized cockpit (see Type Description No. 271 Ta 152 H high-altitude fighter aircraft). A GM 1 system in the fuselage was standard equipment for increased performance above 8,000 meters. Operating duration was approximately 17 minutes at an average consumption of 100 g/sec.

Prototypes, Production and Variants of the Ta 152 E

Originally only one prototype was planned for the reconnaissance versions of the Ta 152, however this was subsequently changed to three prototypes. The following reconnaissance prototypes of the Ta 152 were planned:

Ta 152 V 9
Werknummer 110 009
Prototype for the Ta 152 E-1
Ta 152 V 14
Werknummer 110 014
Prototype for the Ta 152 E-1
Ta 152 V 26
Werknummer 110 026
Prototype for the Ta 152 H-10

The Ta 152 V9 and V14 prototypes, which were originally earmarked for testing the Ta 152 E-1 and which were supposed to be ready to fly on 18 January and 25 January 1945 respectively, were dropped on 5 January 1945 after discussions with the RLM. After this the first installation of standard equipment was to be made in the first production Ta 152 E aircraft in January 1945. MMW was supposed to begin production in February 1945 and plans called for the periscope to be dispensed with. Work began on a batch of thirty reconnaissance machines. Type inspection of the first E-1 did not take place until 1 March 1945. Center of gravity problems were anticipated as a result of the weight of the cameras (approx. 70 kg). It is uncertain whether this reconnaissance aircraft left the factory. The military situation forced a change in the program. MMW was now ordered to take part in production of the Ta 152 C and the E-fuselage was to be used as a component for the Ta 152 C. This version received the new designation Ta 152 C-11.

The RLM also called for a version with an oblique camera installation (Ta 152 E-1/R1), whereby the RB 75/30 camera was replaced with a RB 50/18 mounted at 10 degrees below the horizontal. A bulge on the side of the fuselage was necessary in order to install the camera obliquely in the air-

frame of the Ta 152, and this required aerodynamic investigations by Focke-Wulf. Consequently the side bulge for the oblique camera installation was tested on Fw 190 D-9 *Werknummer* 210 002, TR + SB. The pilot of the Ta 152 E-1/R1 would have to roll the aircraft onto its side in order to take pictures, similar to the procedure used by some reconnaissance versions of the Spitfire. The escort reconnaissance machine demanded by the OKL was built on the basis of the Ta 152 H and displayed the same changes as on the Ta 152 E. Originally designated the Ta 152 E-2, twenty examples of the Ta 152 H-10 reconnaissance aircraft were to be built per month. The series prototype was to be the Ta 152 V 26, a standard Ta 152 H-0 or H-1 converted to Ta 152 H-10 standard by MMW. It is very likely that WNr. 150 167 was the aircraft selected for conversion, for this Ta 152 was captured at Erfurt. MMW was scheduled to begin production in May 1945. Construction of the Ta 152 H-10 was basically similar to that of the Ta 152 H-1, with the MW 50 wing bag tank also capable of being switched to B4 fuel. A 300-liter drop tank could also be carried.

Another front view of 150 167.

125

The Planned Test Series from Sorau

Focke-Wulf tried to investigate potential problems associated with production of the Ta 152 as early as possible by initiating a very large prototype construction program. Therefore, in the summer of 1944 a prototype program for an initial total of 26 test-beds was created; the aircraft were to be new machines built in the Sorau[1] production facility. Construction of the originally planned prototypes for the Ta 152 A-1, the Ta 152 V 1 and V 2, and for the Ta 152 H-1, the Ta 152 V 3-V 5, was dropped early on. In order to gain time for testing, Focke-Wulf decided to create the prototypes for the high-altitude fighter from four non-production converted aircraft at Adelheide (the Fw 190 V 33/U1, Fw 190 V 30/U1, Fw 190 V 29/U 1 and Fw 190 V 18/U2). As early as 18 July 1944 *Oberst* Petersen, commander of the Rechlin Test Station, made reference to the difficulties of the current testing situation. Concerning the Ta 152 H he remarked: "Since it is planned to start production of the Ta 152 H with only four non-series prototypes[2] and thus with insufficient testing, delays must be expected for the series as a result of complaints about flight safety and flight characteristics as well as a considerable number of initial modifications. In any case the first twelve production aircraft will be needed for testing (reference is made to roughness in the Jumo 213 E power plant)." Petersen also referred to the beginning of large-scale production

Overview of Ta 152 Prototypes from Sorau/Adelheide (Production Block 110)

V-No.	WNr.	Registration	Remarks
V 1	250 001		Ta 152 A-1, not built
V 2	250 002		Ta 152 A-1, not built
V 3	260 001		Ta 152 H-1, not built
V 4	260 002		Ta 152 H-1, not built
V 5	260 003		Ta 152 H-1, not built
V 6	110 006	VH + EY	Ta 152 C-0, first flight on 12/12/1944, pilot Märschel. DB 603 E, V 19 prototype engine, engine no. 0130 0145
V 7	110 007	CI + XM	Ta 152 C-0/R11, first flight on 08/01/45, pilot Märschel. DB 603 E, V 20 prototype engine, engine no. 0130 0147.
V 8	110 008	GW + QA	Ta 152 C-0/EZ, first flight on 15/01/45, pilot Märschel. DB 603 E, V 21 prototype engine, engine no. 0130 0150.
V 9	110 009		Ta 152 E-1, canceled 05/01/45
V 10	110 010		Ta 152 C-1, canceled 18/10/44
V 11	110 011		Ta 152 C-1, canceled 18/10/44
V 12	110 012		Ta 152 C-1, canceled 18/10/44
V 13	110 013		Ta 152 E-1, planned ready to fly date: 25/12/44. After 19/10/44 as Ta 152 C-1/R11 with LGW K 23 autopilot, new planned ready to fly date: 06/02/45.
V 14	110 014		Ta 152 E-1, cancelled 05/01/45.
V 15	110 015		Ta 152 C-2/R11 with LGW K23 autopilot instead of PKS 12, new planned ready to fly date: 14/02/45.
V 16	110 016		Ta 152 C-3, planned: DB 603 LA/MW 50, planned ready to fly date approx. March 45.
V 17	110 017		Ta 152 C-3, planned: DB 603 LA/MW 50, planned ready to fly date approx. March 45.
V 18	110 018		Ta 152 C-3, planned: DB 603 LA/MW 50, then Ta 152 C-4/R11, canceled 28/12/44.
V 19	110 019		Initially Ta 152 C-5, then Ta 152 B-5, planned ready to fly date: April 1945.
V 20	110 020		Initially Ta 152 C-5, then Ta 152 B-5, planned ready to fly date: May 1945.
V 21	110 021		Initially Ta 152 C-5, then Ta 152 B-5, planned ready to fly date: May 1945.
V 22	110 022		Ta 152 C-4, canceled 18/10/44.
V 23	110 023		Ta 152 C-4, canceled 18/10/44.
V 24	110 024		Ta 152 C-4, canceled 18/10/44.
V 25	110 025		Ta 152 H-1, construction of prototype halted, completed wing with 4-tank installation used for Fw 190 V 32/U1.
V 26	110 026		Ta 152 H-10, reordered as converted Ta 152 H-0, probably WNr. 150 167. Planned ready to fly date: March 1945.
V 27	150 027		Converted Ta 152 H-0 for testing MK 103 cannon for Ta 152 C-3, DB 603 E engine.
V 28	150 028		Converted Ta 152 H-0 for testing MK 103 cannon for Ta 152 C-3, DB 603 E engine. Planned ready to fly date: 18/02/45.

of the Ta 152 C and expressed the opinion that in this case too 30 aircraft would have to be held back for trials. Since there was as yet no test data available for the DB 603 L engine, which was to power the Ta 152 C series, the planned January 1945 start-up date for production of the Ta 152 C with the DB 603 L was seen as uncertain.

The subsequent production plans for the Ta 152 C prototypes could also not be held to on account of the war situation and it is questionable whether more than three new-build prototypes of the Ta 152 C-0 (V 6, V 7, V 8) and one converted aircraft (Ta 152 V 27 or 28) were ever built and flown. Here, too, the military situation forced pro-

[1] Sorau today is called Silesia and is in Poland.
[2] The Fw 190 V 33/U1, V 30/U1, V 18/U2 and V 29 U1.

totype construction to be moved to Adelheide. The prototypes Ta 152 V 27 and V 28 were an exception.

In order to test the installation of the engine-mounted ML 103 cannon in the DB 603 power plant for the production Ta 152 C-3, it was planned to convert two production Ta 152 H-0 (*Werknummer* 150 027 and 150 030) in February 1945. The combination of DB 603 E engine and MK 103 cannon was successfully tested in at least one of these aircraft. It may also be assumed that construction of the Ta 152 V 16, V 17, V 19, V 20 and V 21 prototypes was at an advanced stage.

Whether any of these aircraft flew cannot be determined with certainty. According to statements by chief test pilot Sander these prototypes were not completed.

The High-Altitude Power Plants for the Ta 152

It was planned to install the new Junkers 9-8213 FH power plant, which became the Jumo 213 E, in the Ta 152 H high-altitude fighter. The Jumo 213 E was a 12-cylinder inline engine with a mechanically-driven two-stage supercharger and three-speed transmission. The Jumo 213 E produced 1,730 H.P. (1,272 kW) for takeoff at 3,250 rpm. Even at its maximum boost altitude of 10,700 meters the Jumo 213 E still produced 1,260 H.P. (927 kW) without injection boosting. In order to avoid thermal overheating the engine required an intercooler, and in contrast to the Fw 190 D this could be accommodated in the airframe of the Ta 152 H. At the same time output could be increased further through the injection of MW 50 and GM 1. Using MW 50 the Ta 152 H-1 could reach 749 kph at an altitude of 9,500 meters and with GM 1,760 kph at 12,500 meters. But in order to achieve these performances the engine had to be made fully ready for production.

But as a result of extraordinary time constraints the first Jumo 213 Es were installed in the first Ta 152 H fighters. With the small number of hours logged by the Ta 152 H prototypes and the engine failures encountered in testing, it proved impossible to make the engine fully production-ready. As a result, when the Ta 152 H entered production in October-November 1944 emergency power could not be used in the high range (3rd gear) because of weakness in the transmission. Not until the improved Jumo 213 E-1 was available could emergency power again be used. The second handicap faced by the Ta 152 H-0 series was the absence of power boosting systems. Junkers subsequently made the proposal that the machines' performance could be increased

by installing an equipment set (*Rüstsatz*) which raised boost pressure. This equipment set, which was also successfully installed in all Fw 190 D-9s, increased engine output by 150 H.P. to 1,900 H.P. In February 1945 Jumo's technical field service began converting the then Ta 152 H-0s still under test by III/JG 301. After flying the first converted aircraft the pilots declared themselves satisfied with the increased performance of the Ta 152 H.

Overhead view of the Jumo 213 E high-altitude engine, here installed in the Ta 152 H. The engine's early development was troublesome.

129

Testing the DB 603 LA, which was earmarked for the Ta 152 C and E series proved even more difficult. Like the Jumo 213 F, which powered the Fw 190 D-11, 12 and 13, the DB 603 LA's supercharger air was cooled by MW 50, since a supercharger intercooler was not initially planned. Not until the later DB 603 L did an intercooler appear. Without large-scale testing, beginning in March 1945 the DB 603 LA engine was to be installed in the production Ta 152 C.

All of this shows that a proven, reliable high-altitude engine was a necessity for the Ta 152 high-altitude fighter. But as the engine modifications that were undertaken show, the Jumo 213 E did not achieve a satisfactory level of operational reliability until midway through the series production of the Ta 152.

Equipping the Ta 152 with Special Weapons

In the last weeks of February 1945 an entire *Staffel* of Fw 190 D-9s belonging to the test unit *Jagdgruppe* 10 at Parchim was equipped with R4M air-to-air rockets.[1] Each Fw 190 D-9 was equipped with two underwing launch racks, each with twelve rails, carrying a total of twenty-four R4M. JGr. 10 had been specially assigned to test the R4M. The R4M was the only air-to-air rocket to see service with the *Luftwaffe* and its warhead contained a high-explosive charge weighing 540 grams (R=*Rakete*, 4.4 kg=weight, M=*Minenkopf*).

As a result of these trials the order was issued to immediately equip not just the Me 262 but also the Fw 190 and Ta 152 with the R4M. The R4M installation was to be installed on the Ta 152 C-1/R31 and the Ta 152 H-1/R31. Modification directives also anticipated retrofitting the Ta 152 H-1/R11, H-1/R21 and C-1/R11. By the end of the war JGr. 10 was to have three Ta 152s in addition to the rocket-armed Fw 190s. The effectiveness of the R4M was successfully demonstrated by the Me 262. The order that disbanded *Jagdgruppe* 10 on 2 April 1945 called for all of 2/JGr. 10's Fw 190 D-9s equipped with the R4M to be handed over to I/JG 301; it is not known whether this order was carried out.

Another special weapon planned for use by the Ta 152 was the SG 500 (*Sondergerät* 500). This recoilless weapon was designated the "*Jagdfaust*." Five units were to be installed vertically in each wing of the Ta 152, resulting in the elimination of one fuel tank in each wing. The weapon itself consisted of a firing tube 0.515 meters long that weighed 6.9 kg. The projectiles used were spin-stabilized special munitions with a caliber of 50 mm. Release was photo-electric by means of a selenium cell which fired the SG 500 "*Jagdfaust*" when the fighter flew under an enemy bomber. The Tarnewitz Test Station conducted successful trials with the weapon installed in a Fw 190, however the end of the war precluded its production and use.

A similar fate was suffered by another special weapon, the R*ohrblock* 108. This weapon was also recoilless and two units were supposed to be installed vertically in each wing. The *Rohrblock* consisted of seven MK 108 cannon and was also supposed to be fired photo-electrically while passing beneath an enemy bomber.

The Planned Ta 152 S Two-Seat Trainer

The Ta 152 S two-seat trainer was based on the Ta 152 C-1 powered by the DB 603 and was to be fitted with a second cockpit with instrumentation. The fuselage conversions were supposed to be carried out by Blohm & Voss beginning in April 1945 and DLH (Deutsche Lufthansa) in Prague beginning in August 1945. Although by this stage of the war the shooting down of German training aircraft had become the order of the day, no armament was planned for the Ta 152 S-1.

Conversion of the Ta 152 into a two-seat trainer was supposed to follow the pattern of the Fw 190 two-seater (Fw 190 S-5, S-8). The first Ta 152 training aircraft was originally planned for November 1944, but

[1] War diary, chief Technical Air Armaments, Flight Test Working Group from 26 February to 4 March 1945.

131

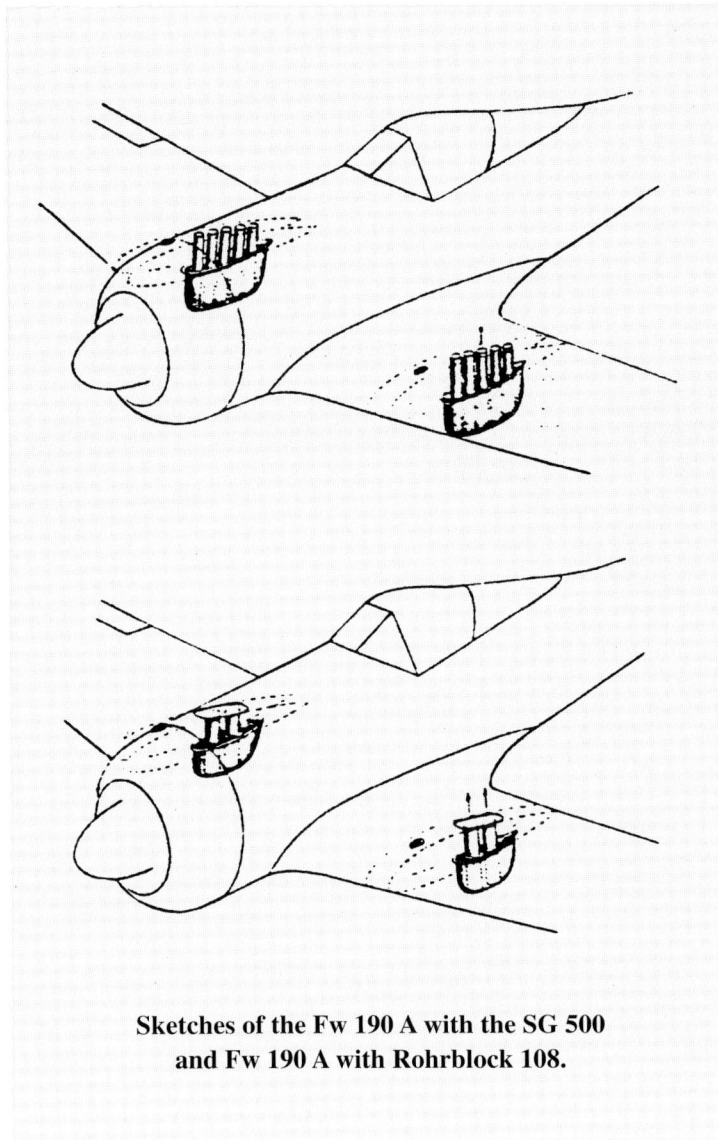

**Sketches of the Fw 190 A with the SG 500
and Fw 190 A with Rohrblock 108.**

testing Fw 190 D test-beds Focke-Wulf discovered that merely sealing the engine cowling joint lines with rubber resulted in a speed increase of up to 17 kilometers per hour for the Fw 190 D-9 and D-11. This of course also applied to the Ta 152. Consequently greater attention was to be paid to sealing during construction.

Another possible way of increasing performance was raise the boost pressure of the Jumo 213. In March 1945 similar modifications to those made to the Jumo 213 A installed in the Fw 190 D were carried out on the ten Ta 152 H-0s of III/JG 301, resulting in an increase in engine takeoff power to 1,900 H.P. (1 397 kW).

An entirely new method of increasing the Ta 152's performance was the use of the so-called integral engine cowling. It was anticipated for the Ta 152 and the Fw 190 D-12. For the Ta 152 C series the prototypes Ta 152 V 6 and V 7 were converted with the integral engine cowlings and were supposed to be ready to fly on 28 February and 10 March 1945 respectively. Later additional test-beds were to include the Ta 152 V 16, V 17, V 27 and V 28. The integral engine cowling was even planned for the already potent Ta 152 H. A total of twenty prototypes were supposed to be built or converted to test the integral cowling on the Ta 152 H. To come first, however, were the Fw 190 V 32/U2, Ta 152 V 19, Ta 152 V 20, Ta 152 V 21 and the Ta 152 H WNr. 150 004. Junkers expected to begin production of Jumo 213 engines with the integral cowling by August 1945 at the earliest. No definitive decision as to whether the Jumo 213 E-1 should be so modified could be reached by March 1945.

Two versions of the integral engine cowling were planned. Version 1 saw the installation of vibration elements between the engine and the integral cowling. This version was too costly for mass production and was therefore not to be adopted for large-scale use.

Only one Ta 152 H-0, *Werknummer* 150 004, was equipped with this cowling. In the process it may have been built faster

the serious production delays that affected the Ta 152 C frustrated these plans. The Rechlin Test Station demanded that three percent of all production Ta 152 Cs be trainers. A total of 565 Ta 152 S-1s were planned by March 1946, none were built. The Ta 152 S-2 proceeded no farther than the design stage and was not pursued.

**Improving the Performance
of the Ta 152**

Toward the end of the war the Focke-Wulf development section busied itself with various possibilities of enhancing the performance of the Ta 152 and Fw 190 D. While

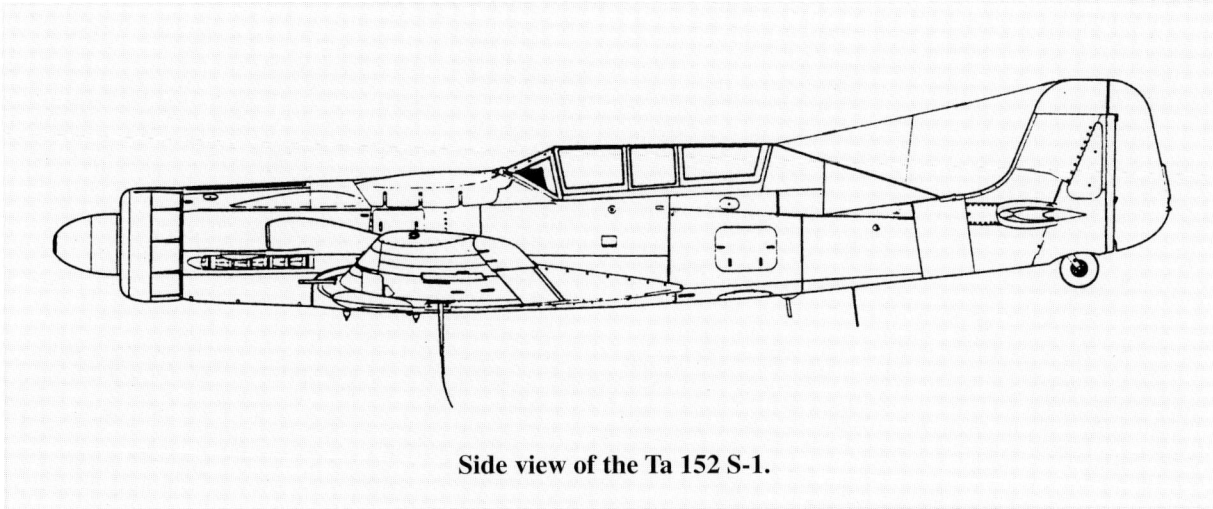

Side view of the Ta 152 S-1.

than any other Ta 152 H. The simplified Version 2 saw the installation of a roller chain. Version II was supposed to be installed on all production Ta 152s.

No detailed test reports have so far been discovered, although the Ta 152 V 6 (VH + EY) and Ta 152 V 7 (CI + XM) were converted to this integral engine cowling and achieved satisfactory results during trials. The hopeless war situation prevented quantity production.

On 12 March 1945 Focke-Wulf issued its last summary of the prototypes on hand at Langenhagen and their test programs. The following types were on hand: four Ta 152s, two Ta 152 Cs, two Fw 190 D-9s, two Fw 190 D-11s and four other Fw 190s. According to the summary the following test program was foreseen for the Ta 152 (priority level 1):

Ta 152 H V 29
Pressurized cockpit testing, new AR 300 chamber regulator.

V 18
Integral engine cowling, wooden tail.

150 004
Handling characteristics with additional fuselage inserts of 300 mm and 500 mm as well as enlarged horizontal tail and modified wing-fuselage junction.

V 32
Fuel and methanol wing tanks, integral engine cowling, engine trials Jumo 213 E-1.

Ta 152 C V 6
Power plant trials with DB 603 LA. Handling characteristics after lowering engine, performance measuring, integral engine cowling.

Ta 152 C V 7
Power plant trials with DB 603 LA, integral engine cowling. Handling characteristics after installation of torpedo, performance measuring.

Chief test pilot Hans Sander recalls that there was already a terrible fuel shortage at that time. Only one third of the amount required by the test aircraft was received. The two-year test period at Langenhagen was coming to its bitter end.

At this time Focke-Wulf's last design work on the Ta 152 was taking place in Bad Eilsen. In addition to aerodynamic improvements, preliminary investigations for the installation of the Jumo 213 J were carried out and armor for use by Ta 152 *Sturmstaffeln* was developed. Other pending development tasks for the Ta 152 were: use of MW 100, sliding canopy using silicate glass and design of an integral upper engine cowling without gun channels. A prototype

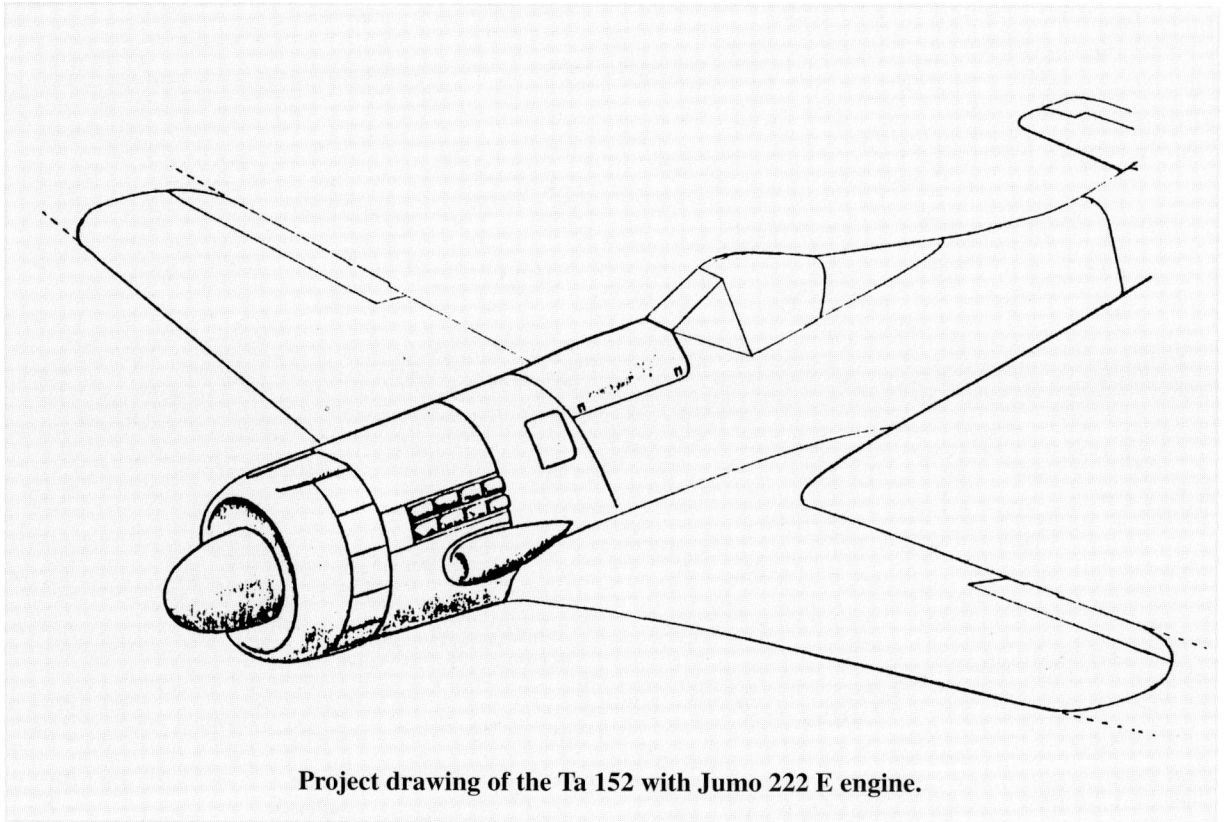

Project drawing of the Ta 152 with Jumo 222 E engine.

installation of the new Mauser MG 213 cannon in the Ta 152 was also to be carried out.

Further Development of the Ta 152 H with the Jumo 222

At the end of 1944 Focke-Wulf was still engaged in further development of the Ta 152 H. The Ta 152 H was to be equipped with the Jumo 222 E[25] and receive a new wing with a laminar profile. The Jumo 222 E could be exchanged for the Jumo 222 A at any time. All performance figures and requirements were laid down in the brief description subsequently issued by Focke-Wulf. But by this time it must have already been clear that the Ta 152 with Jumo 222 could only remain a pure development project. It would also have been questionable if this development would have been realized in view of the obvious superiority of the jet fighter and the development of Focke-Wulf's own jet design, the Ta 183.

The Jumo 222 engine planned for installation in the Ta 152 had already played a decisive role in the ill-fated "Bomber-B Program."[26] Earmarked for the Ju 288, successor to the Ju 88, the engine was never put into large-scale production. The Jumo 222 was a liquid-cooled, 24-cylinder radial engine with a displacement of 49.85 liters and produced 2,900 H.P. for takeoff with MW 50 injection. The Jumo 222 E planned for the Ta 152 H was supposed to drive a Junkers VS 19 propeller (four-bladed propeller with wooden blades, diameter 3.60 m). The new wing was designed as a monocoque structure with two spars and had a laminar profile of d/ti=15.85 percent and d/ta=10 percent.

[1] Brief description No. 25 dated 4 December 1944.
[2] Procurement program for a medium bomber 1940-41.

134

Taking into consideration the latest aerodynamic knowledge, a laminar profile was incorporated into the root profile which was retained as far as the undercarriage attachment points.

As the weight estimates show, Focke-Wulf calculated that, like the Ta 152 H with the Jumo 213 E, the new wing with its six fuel tanks would not be available in time. Therefore balance calculations were made for an aircraft without wing tanks and MW 50 and one with wing tanks and MW 50 injection.

Type Sheet Ta 152 with Jumo 222 E

Purpose:	Single-seat fighter, fighter-bomber
Configuration:	Single-engined cantilever monoplane with hydraulically-retractable undercarriage
Power plant:	Jumo 222 E with methanol-water (MW 50) injection
Dimensions:	Wing area: 23.8 m²
	Wingspan: 13.68 m
	Aspect ratio: 7.9
	Vertical tail area: 1.78 m²
	Horizontal tail area: 2.89 m²
	Maximum length: 10.77 m
	Maximum height: 3.75 m
	Normal takeoff weight: 5 815 kg
Armament:	2 MG 151/20 in fuselage with 150 rounds per gun
	2 MG 151/20 in wing roots with 175 rounds per gun
	or
	2 MG 151 in fuselage with 150 rounds per gun
	2 MK 103 in wing roots with 55 rounds per gun
Armor:	Engine armor in front of firewall: 76 kg
	Fuselage armor: 81 kg
	Total weight of armor: 157 kg
Equipment:	FuG 16 ZY, FuG 25a, FuG 125, K 23 automatic pilot, Revi 16b
Fuel system:	Ta 152 with Jumo 222
	232 l in forward tank
	360 l in rear tank
	240 l in left wing
	240 l in right wing
Performance:	takeoff power with Jumo 222 E: 2,500 H.P. at 3,000 rpm
Maximum speed:	710 kph at 9 500 m—emergency boost with MW 50
Service ceiling:	15 000 m
Range:	1 290 km at 10 000 m without 300-l drop tank
Rate of climb at ground level:	22 m/sec.

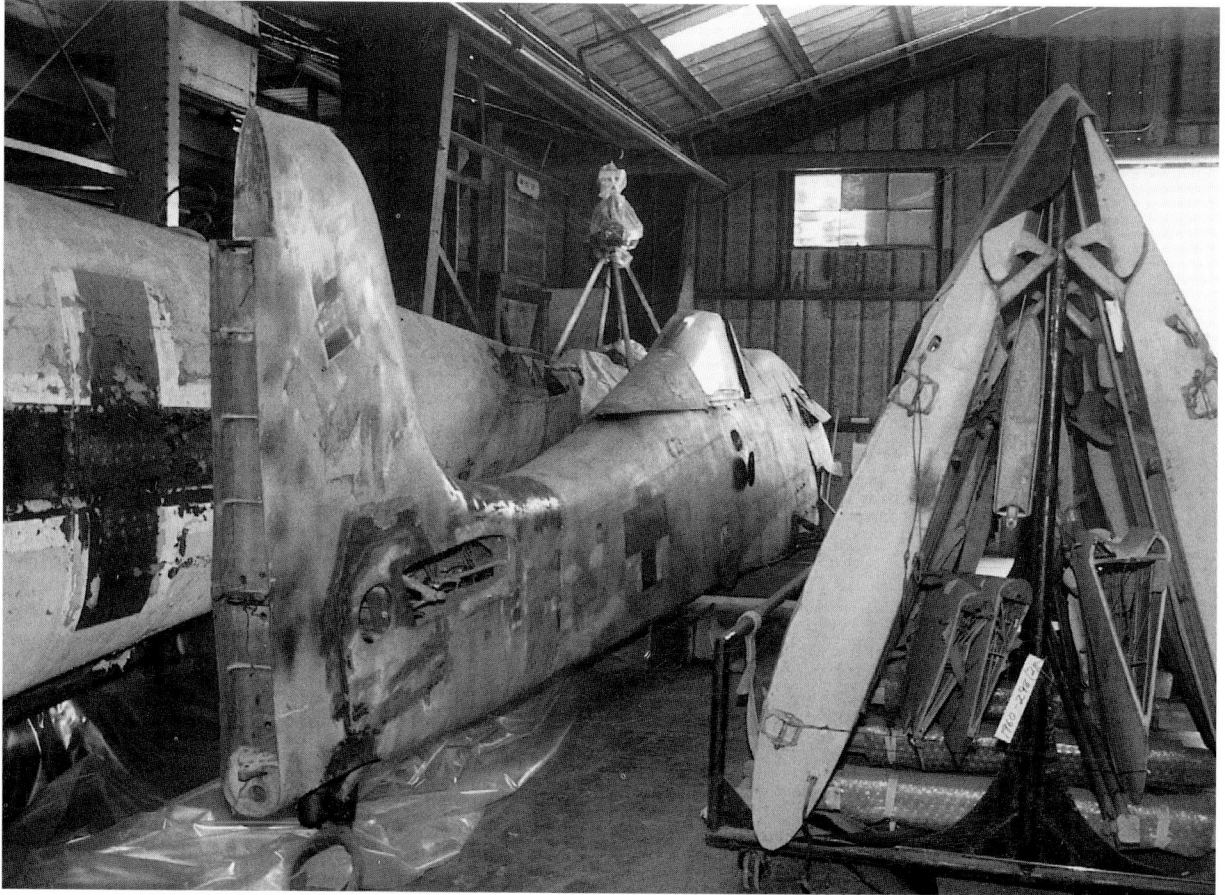

Ta 152 H-0, *Werk-nummer* 150 010, in storage in the NASM's Silver Hill depot in Washington.

Another view of the stored Ta 152 H-0, which is not scheduled to be restored in the foreseeable future.

Einmotorige Jäger: Leistungsdaten

	Fw190A-8	Fw190A-9	Fw190D-9	Fw190D-12	Ta152H-0	Ta152C-0	Ta152E-0	Fw190(D-9)	Fw190(D-9)	Ta152(C-0)
Motormuster	BMW801D	BMW801F	Jumo213A	Jumo213F	Jumo213E	DB603L	Jumo213E	DB603A	DB603E	DB603E
Bewaffnung: Motor	–	–	–	1×MK108	1×MK108	1×MK108	1×MK108	–	–	1×MK108
Bewaffnung: Rumpf	2×MG131	2×MG131	2×MG131	–	2×MG151	2×MG151	2×MG151	2×MG131	2×MG131	2×MG151
Bewaffnung: Flügel innen	2×MG151	2×MG151	2×MG151	2×MG151	–	2×MG151	–	2×MG151	2×MG151	2×MG151
Bewaffnung: Flügel außen	2×MG151	2×MG151	–	–	–	–	–	–	–	–
Abfluggewicht (kg)	4 300	4 370	4 300	4 420	4 760	4 830	4 675	4 400	4 400	4 770
Höchstgeschwindigkeit mit Notleistung am Boden (km/h)	548 (578)	560	576 (612)	572 (608)	540 (580)	(576)	547 (581)	556 (607)	563 (607)	543 (590)
″ (km/h)	644 (652)	666	685 (702)	738 (738)	720 (742)	(753)	712 (744)	672 (691)	687 (706)	667 (684)
in Volldruckhöhe (km)	6,3 (5,5)	6,4	6,6 (5,7)	11,6 (11,6)	10,7 (9,5)	(10,5)	10,7 9,5	7,0 (5,1)	8,3 (6,9)	8,2 (6,8)
Höchstgeschwindigkeit mit Kampfleistung (km/h)	614	638	675	710	704	(710)	695	660	675	654
in Volldruckhöhe (km)	5,8	6,1	6,6	10,9	10,5	(10,5)	10,5	6,9	8,3	8,2
Steigleistungen m. Kampfleistung: Dienstgipfelhöhe w_{st}=0,5 m/s (km)	9,95 (10,6)	10,8	11,1 (11,6)	12,8 (13,4)	13,5 (15,5)× (13,9)	(13,0)	12,4 (12,9)	11,0 (11,4)	11,6 (12,0)	11,1 (11,5)
Arbeitshöhe w_{st}=2,0 m/s (km)	9,25 (9,9)	10,1	10,4 (11,0)	12,3 (13,0)	12,9 (14,4)× (13,3)	(12,5)	11,9 (12,4)	10,4 (10,8)	10,7 (11,4)	10,5 (10,9)
Steiggeschwindigkeit (m/sec)	9,6 (14,0)	11,7	12,7 (18,5)	9,0 (8,2)	9,7 (14,5)	(10,9)	7,6 (13,0)	12,1 (19,6)	10,7 (16,5)	9,1 (14,6)
in Volldruckhöhe (km)	5,5 (4,9)	5,75	5,8 (4,8)	10,3 (11,2)	9,9 (8,8)	(9,5)	9,9 (8,8)	6,0 (4,0)	7,5 (5,9)	7,5 (5,9)
Steigzeit auf 10 km (min)	19,6	19,6	16,8 (12,5)	13,1 (10,9)	13,8 (10,1)	(10,9)	15,6 (10,9)	17,1 (12,3)	15,0 (11,2)	18,4 (13,6)
Rollstrecke auf Beton (m)	430	390	365	355	295	380	375	420	385	420
Startstrecke bis 20 m (m)	715	600	570	575	495	605	595	630	600	635
Steiggeschwind. b. Abheben (m/s) (Fig. im Startzustand)	9,7	11,5	11,3	12,9	12,0	12,9	12,8	10,6	11,1	10,7
Leistungsverluste bei geänderter Bewaffnung			2 MG151 + 2 MK108		×mit GM1	3×MK103		2MG151+2MK108	2MG151+2MK108	
Δ V_{max} in V.Dr.H. (km/h)			10			18		10	10	
Δ w_{st} (m/sec)			0,45			0,70		0,45	0,45	
Δ H Dienstgipfelhöhe (m)			150			300		150	150	

Fw190A-8: Leistungen gelten nur für Lüfterrad 032 bei 035 ist zusätzlich Leistungsverlust im BL-Ast vorhanden.

Fw190D-12: Staatsgeheimnis! Geheimhaltungspflicht beachten

Geschwindigkeiten ohne Berücksichtigung des Widerstandsanstieges aus Kompressibilität, ohne ETC unter dem Rumpf, mit beweg. Fahrwerks-klappen, Oberfläche gespachtelt u. Glattanstrich.

Eingeklammerte Werte gelten für Sondernotleistung (Start- u. Notleistung mit MW 50)
Bei Fw190 A-8 für Notleistung und erhöhtem Ladedruck!

1.10.1944

Fw 190 – Ta 152
Einmotorige Jagdflugzeuge

		Baureihen	Motor	Bewaffnung I	Bewaffnung II	Fluggewicht bei Bewaffnung II	Flugdauer in 7 km H. u. Sparleistg.
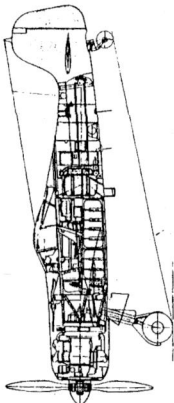 **Fw 190 A**	Jagdflugzeug	A1 ÷ A9	BMW 801D TU;TS;TH	ab A-7	ab A-7/R2		mit 300l Abw.Beh. 3,3ʰ / ohne 2,2ʰ
	Schlachtflugzeug	F1 ÷ F16		ab F-8		4 500 kg	— / 1,9ʰ in 5 km Höhe
	Jagdbomber	G1 ÷ G9		300 ltr.	SC 500	5 050 kg	4,10ʰ in 5 km Höhe / —
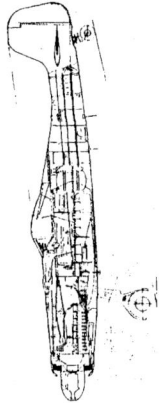 **Fw 190 D**	Jagdflugzeug	D-9 Serie Aug. 44	Jumo 213 A	—		4 300 kg	mit MW50 2,90ʰ / ohne MW50 3,5ʰ 1,7ʰ / 2,2ʰ
	Jagdflugzeug	D-12 Serie Jan. 45 / D-13	Jumo 213 F	D-13	D-12	4 450 kg	Serien-Anlauf 2,7ʰ 1,6 / — 2,6
	Jagdflugzeug	D-11 Serie Jan. 45	Jumo 213 F				
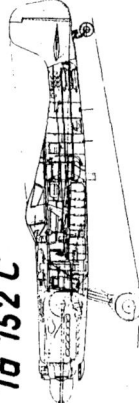 **Ta 152 C**	Jagdflugzeug	C-1 Serie März 45 / C-3	DB603L	C-3	C-1	5 320 kg	Serien-Anlauf mit Druckkabine 3,2ʰ
	Jagdflugzeug	B-5 Serie Mai 45	Jumo 213 E				
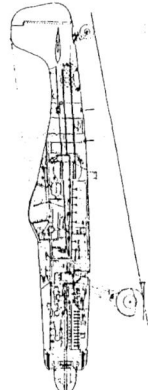 **Ta 152 H**	Begleitjäger	H-1 Serie Jan. 45	Jumo 213 E			5 220 kg	Serien-Anlauf 2,9ʰ / 2,0ʰ Endg.Serie mit Druckkabine 3,3ʰ
	Aufklärer	E-1 Serie Febr. 45				5 085 kg	Serien-Anlauf 2,8ʰ / 1,8ʰ Endg.Serie mit Druckkabine 4,5ʰ / 3,4ʰ
	Höhenaufklärer	H-10 Serie Mai 45				5 280 kg	Serien-Anlauf mit MW50 / Endg.Serie mit Druckkabine 3,3ʰ

Übersicht der Ta 152 Versuchsmuster

Serie	Versuchsmuster	Kennung	Werknummer	Erprobung	Motor	Erstflug
Ta 152 A	Fw 190 V 19	—+—	0041	Fahrwerkshydraulik, MK 103 Einbau Fl.V.Anlage geplant	Jumo 213 C	07.07.1943
	Fw 190 V 20	TI+IG	0042	Fl.V.Anlage, Einheitstriebwerk, Hydraulik	Jumo 213 C	23.11.1943
	Fw 190 V 21	TI+IH	0043	Fl.V.Anlage, Einheitstriebwerk, Hydraulik	Jumo 213 C	13.03.1944
Ta 152 B	Fw 190 V 68	DU+JC	170003	Waffenerprobungsträger MK 103 im Flügel	Jumo 213 A	13.12.1944
Ta 152 C	Fw 190 V21/U1	TI+IH	0043	Motorerprobungsträger	DB 603 E	03.11.1944
	Ta 152 V 6	VH+EY	110006	Ta 152 C-0	DB 603 E	12.12.1944
	Ta 152 V 7	CI+XM	110007	Ta 152 C-0/R 11	DB 603 E	09.01.1945
	Ta 152 V 8	GW+QA	110008	Ta 152 C-0 mit EZ 42[1]	DB 603 E	15.01.1945
	Ta 152 V 27	—+—	150028	Motor-MK 103 Erprobung für Ta 152 C-3	DB 603 E	07.02.1945[3]
	Ta 152 V 28	—+—	150030	Motor-MK 103 Erprobung für Ta 152 C-3	DB 603 E	14.02.1945[3]
Ta 152 E	Fw 190 D-9	TR+SB	210002	Kameraschrägeinbau Ta 152 E-1/R1	Jumo 213 A	15.09.1944[2]
Ta 152 H	Fw 190 V 33/U1	GH+KW	0058	Musterflugzeug ohne MW 50/GM1	Jumo 213 E	13.07.1944
	Fw 190 V 30/U1	GH+KT	0055	Musterflugzeug ohne MW 50/GM1	Jumo 213 E	06.08.1944
	Fw 190 V 29/U1	GH+KS	0054	Musterflugzeug ohne MW 50/GM1	Jumo 213 E	24.09.1944
	Fw 190 V 18/U1	CF+OY	0040	Musterflugzeug ohne MW 50/GM1	Jumo 213 E	19.11.1944
	Fw 190 V 32/U1	GH+KV	0057	Musterflugzeug mit Original H-1 Flügel	Jumo 213 E	unbekannt
	Ta 152 V 25	—+—	110025	Ta 152 H-1, Flügel für Fw 190 V 32/U1	Jumo 213 E	Bau eingestellt

Bemerkung: Die nicht gebauten Versuchsmuster aus dem Versuchsmusterbau Sorau / Adelheide sind hier nicht erwähnt.

1 EZ 42 = automatische Visiereinrichtung
2 Erstflug als Serien - Fw 190 D- 9
3 geplanter FKT

Baureihenübersicht der Ta 152

Baureihe	Motor	Zweck	Spannweite	Fläche	Länge	Position 1	Position 2	Position 3	Position 4	Position 5
Ta 152 A-1	Jumo 213 A	Jäger/Jabo	11,00 m	19,5 m²	10,78 m	MK 108	MG 151	MG 151	Rüstsatz	500 kg / 300 l
Ta 152 A-2	Jumo 213 A	Jäger/Jabo	11,00 m	19,5 m²	10,78 m	MK 103	MG 151	MG 151	Rüstsatz	500 kg / 300 l
Ta 152 B-1	Jumo 213 E	Jäger/Jabo	11,00 m	19,5 m²	10,78 m	MK 108	MG 151	MG 151	Rüstsatz	500 kg / 300 l
Ta 152 B-2	Jumo 213 E	Jäger/Jabo	11,00 m	19,5 m²	10,78 m	MK 103	MG 151	MG 151	Rüstsatz	500 kg / 300 l
Ta 152 B-5	Jumo 213 E	Zerstörer	11,00 m	19,5 m²	10,78 m	MK 103	–	MK 103	–	500 kg / 300 l
Ta 152 C-0	DB 603 E	Jäger/Jabo	11,00 m	19,5 m²	10,80 m	MK 108	MG 151/20	MG 151/20	–	500 kg / 300 l
Ta 152 C-1	DB 603 L/LA	Jäger/Jabo	11,00 m	19,5 m²	10,80 m	MK 108	MG 151/20	MG 151/20	–	500 kg / 300 l
Ta 152 C-1/R14	DB 603 E/LA	Torpedoflugzeug	11,00 m	19,5 m²	10,80 m	MK 108	MG 151/20	MG 151/20	–	Torpedo LT 1B[2]
Ta 152 C-1/R15	DB 603 E/LA	BT 1400 - Träger	11,00 m	19,5 m²	10,80 m	MK 108	MG 151/20	MG 151/20	–	BT 1400
Ta 152 C-2	DB 603 L/LA	Jäger/Jabo	11,00 m	19,5 m²	10,80 m	MK 108	MG 151/20	MG 151/20	–	500 kg / 300 l
Ta 152 C-3	DB 603 L/LA	Jäger/Jabo	11,00 m	19,5 m²	10,80 m	MK 103	MG 151/15	MG 151/15	–	500 kg / 300 l
Ta 152 C-4	DB 603 L/LA	Jäger/Jabo	11,00 m	19,5 m²	10,80 m	MK 103	MG 151/15	MG 151/15	–	500 kg / 300 l
Ta 152 C-5	DB 603 L/LA	Zerstörer	11,00 m	19,5 m²	10,80 m	MK 103	–	MK 103	–	500 kg / 300 l
Ta 152 C-11	DB 603 L/LA	Jäger/Jabo	11,00 m	19,5 m²	10,80 m	MK 108	MG 151/20	MG 151/20	–	500 kg / 300 l
Ta 152 E-1	Jumo 213 E	Normalaufklärer	11,00 m	19,5 m²	10,81 m	MK 108	–	MG 151/20	–	– / 300 l
Ta 152 E-1/R1	Jumo 213 E	Normalaufklärer	11,00 m	19,5 m²	10,81 m	MK 108	–	MG 151/20	–	– / 300 l
Ta 152 E-2	Jumo 213 E	Höhenaufklärer	14,44 m	23,3 m²	10,81 m	MK 108	–	MG 151/20	–	– / 300 l
Ta 152 H-0	Jumo 213 E	Höhenjäger	14,44 m	23,3 m²	10,71 m	MK 108	–	MG 151/20	–	– / 300 l[1]
Ta 152 H-1	Jumo 213 E	Höhenjäger	14,44 m	23,3 m²	10,71 m	MK 108	–	MG 151/20	–	– / 300 l
Ta 152 H-2	Jumo 213 E	Höhenjäger	14,44 m	23,3 m²	10,71 m	MK 108	–	MG 151/20	–	– / 300 l
Ta 152 H-10	Jumo 213 E	Höhenaufklärer	14,44 m	23,3 m²	10,71 m	MK 108	–	MG 151/20	–	– / 300 l
Ta 152 H-11	Jumo 213 E	Höhenaufklärer	14,44 m	23,3 m²	10,71 m	MK 108	–	MG 151/20	–	– / 300 l
Ta 152 H-12	Jumo 213 E	Höhenaufklärer	14,44 m	23,3 m²	10,71 m	MK 108	–	MG 151/20	–	– / 300 l

Position 1: Zentralwaffe durch die hohle Propellernabe feuernd
Position 2: je 2 gesteuerte Waffen im Rumpf über dem Motor
Position 3: je 2 gesteuerte Waffen in den Tragflächenwurzeln
Position 4: je 2 Waffen in den Außenflügeln
Position 5: Abwurfwaffen/Zusatzbehälter

Bemerkung: Ta 152 C/H R11,R21,R31 siehe Variantenübersicht Ta 152 C bzw. Ta 152 H

1 Möglichkeit für LT 1B kurz (780 kg) bzw. LT 1B lang (850 kg)
2 300 l Zusatzbehälter Ta 152 H an Tanklafette

Leistungsvergleich

Baumuster	Fw 190 V 29/U1	Ta 152 H-0	Ta 152 H-1	Ta 152 H-10	Ta 152 B-5	Ta 152 V 6	Ta 152 C-1
Motor	Jumo 213 E	Jumo 213 E	Jumo 213 E	Jumo 213 E	Jumo 213 E	DB 603 E	DB 603 LA
Normal-Fluggewicht	4200 kg	4727 kg	5217 kg	5280 kg	5450 kg	4370 kg	5322 kg
MW 50 / GM 1	nein / nein	nein / nein	ja / ja	ja / ja	ja / nein	ja / nein	ja / nein
Höchstgeschwindigkeit mit Kampffleistung	708 km/h	706 km/h	1	697 km/h	664 km/h	647 km/h	702 km/h
in Höhe	10 700 m	10 700 m	10 700 m	10 700 m	10 700 m	6850 m	9500 m
Höchstgeschwindigkeit mit Notleistung	714 km/h	718 km/h	2	709 km/h	683 km/h	–	–
in Höhe	10 700 m	10 700 m	10 700 m	10 700 m	10 700 m	–	–
Höchstgeschwindigkeit mit MW 50	–	–	732 km/h	727 km/h	710 km/h	687 km/h	736 km/h
in Höhe	–	–	9500 m	9500 m	9500 m	5250 m	10 000 m
Höchstgeschwindigkeit mit GM-1	–	–	755 km/h	746 km/h	–	–	–
in Höhe	–	–	12 500 m	12 500 m	–	–	–
Dienstgipfelhöhe	13 650 m	13 650 m	14 800 m	14 200 m	11 600 m	10 400 m	12 200 m

1 Geschwindigkeit ähnlich Ta 152 H-10
2 Geschwindigkeit ähnlich Ta 152 H-10

Leistungsvergleich mit der englischen Spitfire und der amerikanischen P-51 Mustang

	Supermarine Spitfire Mk. XIV	North American P 51 D Mustang	Focke-Wulf Ta 152 H-0[1]
Motor:	Rolls Royce Griffon 65	RR Packard Merlin V 1650 - 7	Junkers Jumo 213 E
Startleistung:	1132 kW (1540PS)	1096 kW (1490 PS)	1287 kW (1750 PS)
Vmax:	574 km/h in 0 m Höhe	600 km/h in 0 m Höhe	571 km/h 0 m Höhe
V max.:	707 km/h in 8000 m Höhe	703 km/h in 7 640 m Höhe	718 km/h in 10 700 m Höhe
Leergewicht:	2994 kg	3232 kg	3920 kg
Fluggewicht:	4663 kg	5489 kg	4730 kg
Reichweite:	845 km	1529 km	885 km
Steigrate am Boden:	23 m/s	18 m/s	20 m/s
Steigzeit auf 6 km:	7 min	7 min 18 sec	8 min[2]
durchschn. Steigrate:	14,28 m/s	13,92 m/s	14.58 m/s
Gipfelhöhe:	13 600 m	12 500 m	13 650 m
Spannweite:	11,24 m	11,28 m	14,82 m
Länge:	9,96 m	9,82 m	10,82 m
Höhe:	3,88 m	4,16 m	3,36 m
Bewaffnung:	2 x 20 mm MK (2 x 120 Schuß)	4 o. 6 x 12,7 mm MG Browning 53-2	2 x 20 mm MK 151/20 (2 x 175 Schuß)
	2 x 12,7 mm MG (2 x 250 Schuß)	2 x 400 Schuß + 2 o. 4 x 270 Schuß	1 x 30 mm MK 108 (1 x 90 Schuß)

1 Ta 152 H-0 verfügte weder über MW 50 noch GM 1 - Einspritzung
2 Steigzeit auf 7000 m

Bibliography

Type Descriptions

No. 270 fighter aircraft Ta 152 A and B with standard power plant Jumo 213 A, Jumo 213 E or DB 603 G of 16/12/1943

No. 271 high-altitude fighter aircraft Ta 152 H of 18/01/1944

No. 282 reconnaissance aircraft Ta 152 E-1 and high-altitude reconnaissance aircraft Ta 152 E-2 of 12/07/1944

No. 290 fighter aircraft Ta 152 C with DB 603 LA or DB 603 L of 15/01/1945

No. 292 escort fighter Ta 152 H with Jumo 213 E of 15/01/1945

Brief description No. 25 of 18/12/1944—fighter aircraft Ta 152 with Jumo 222 and laminar wing

Programs

Prototype testing program, date set: 18/08/1944
Fw 190/Ta 152 production program of 05/01/1945
Fw 190/Ta 152 production program of 21/03/1945
Overviews (of the prototypes currently under test)
Overview with date set: 18/07/1943
Overview with date set: 18/09/1943
Overview with date set: 18/10/1943
Overview with date set: 18/12/1943
Overview with date set: 18/10/1944

Development Reports Ta 152 B

Sheet XVII e1, e2 of 22/12/44 "Ta 152 B-5"

Development Reports Ta 152 C

Sheet XVII a1-a3 of 24/08/1944
Sheet XVII a1 of 26/01/1945
Sheet XVII c1, c2 of 16/11/44 "Ta 152 C-3/C-4 with MK 103"
Sheet XVII c1, c2 of 28/12/44 "Ta 152 C-3/C-4 with MK 103"
Sheet XVII d1, d2 of 21/11/1945 "Ta 152 C with synchronized MK 103"

Sheet XVII c3, c4 of 16/01/1945 "Ta 152 C-3 with MK 103 engine-mounted cannon"
Sheet XIX c1, c2 of 21/11/1944 "Bad weather fighter equipment Ta 152 C"
Sheet XIX f1-f3 dated 13/03/1945 "Integral engine cowling"
Comparison of torpedo aircraft Fw 190 F, D and Ta 152 C/R14 of 30/12/1944

Development Reports Ta 152 E

Sheet XVIII a1-a4 of 25/11/1944
Sheet XVIII a1-a4 of 24/01/1945

Development Reports Ta 152 H

Sheet XVI a1-a4 of 23/08/1944
Sheet XVI a1, a2 of 29/12/1944
Sheet XIX c1, c2 of 3/11/1944 "Bad weather fighter equipment Ta 152 H"
Sheet XVI c1-c4 of 13/02/1945 "Special materials systems and measures to improve stability"
Sheet XIX f1 f3 of 13/03/1945 "Integral engine cowling"

Ta 152 A/B

Flight report No. 1 on the Fw 190 / 0042 / V 20 of 28/03/1944
Flight report No. 3 on the Fw 190 / 0042 / V 20 of 25/06/1944
Fw 190 V 20, field trip report by Jumo (Pohle) of 21/11/43 – 24/11/43
Flight report No. 1 on the Fw 190 / 0043 / V 21 of 09/05/1943

Ta 152 C

Test report Ta 152 V 6, V 7 of 27/12/1944 (Daimler Benz)
Firing test with the Ta 152 V 6 to test the fuselage weapons of 30/12/1944
Test report Ta 152 V 6, V 7, V 8 of 14/01/1945 (Daimler Benz)

Flugbericht Nr.1 der Ta 152 V 6 / 110 006 vom
10.02.1945

Triebwerkumstellungen für Ta 152 C vom 9.03.1945

Ta 152 H

Bau eines Musterflugzeuges vom 16.12.1943

Bericht vom Einflugbetrieb in Focke-Wulf Werk
Cottbus im Monat Dezember 1944

OKL-Meldung vom 18.1.1945 über Tiefangriff auf
den Einflughafen Neuhausen am 16.1.1945

Flugversuche zum Nachweis der Flugsicherheit vom
24.01.1945 und vom 19.02.1945

1. Erprobungsbericht Ta 152 H mit Jumo 213 E
Stand vom 30.01.1945

5. Bericht über Erprobung der Ta 152 H-0, Stand vom
14.02.1945

Triebwerkumstellungen für Ta 152 H vom 9.03.1945

Konstruktionsbeanstandungen Ta 152 Meldung Nr. 1/
45 vom 13.03.1945

Bericht des Rechliner Büros über die Erprobung der
Ta 152 H vom 16.03.1945

Junkers-Monatsbericht Gruppe Luftwaffe Februar
1945 vom 16.03.1945

Leistungsvergleich Ta 152 H mit Do 335 Höhenjäger
(Dornier)

Motorenausfälle Jumo 213 E bei Focke-Wulf (Jumo
213) vom 14.10.44

Motorenausfälle Jumo 213 E bei Focke-Wulf (Jumo
213) vom 13.01.45

Bedienvorschrift Fl 8-152 H-0 Teil 1 - Bedienungs-
karte für den Flugzeugführer Januar 1945

Besprechungsniederschriften

Ta 152 H und Fw 190 D-11 (Bad Eilsen) am
22.8.1944

Ta 152 C mit Triebwerk DB 603 E/LA (bei Fw in Bad
Eilsen) vom 13.1.1945

Niederschrift Nr. 5105 Triebwerk 8603 B1/TLE bzw.
TLA für Ta 152 C (Daimler Benz) vom 15.12.1944

Niederschrift Nr. 5112 Triebwerk 8603 B1/TLE bzw.
TLA für Ta 152 C (Daimler Benz) vom 12.02.1945

Flugbeanstandungen Ta 152 vom 15.03.1945

Ta 152 allgemein

1. Technische Bemerkungen v. K.d.E. der Luftwaffe
(Rechlin) vom 7.08.1944

2. Terminablauf 8-152, Chef TLR vom 27.12.1944

3. Varianten der neu anlaufenden Baumuster vom
11.01.1945

4. Lagebericht 8-152, TLR vom 21.02.1945

5. FW-Überblick über die Flugkraftstofflage für
Monat März 1945 vom 12.03.1945

6. Baureihenbezeichnung Fw 190 und Ta 152 vom
15.03.1945

7. Stellungnahme zum Schreiben des G.d.J. vom
17.03.1945 betr. 8-152 H u. C, Stabilitätsschwierig-
keiten v. 29.03.1945

Bücher/Zeitschriften

National Air & Space Museum Vol. 9 – FW 190,
Workhorse of the Luftwaffe

William Green – Famous Fighter of the II. WW

Gert W. Heumann FR 7/65 – Focke-Wulf Fw 190 Teil
2: Die Entwicklung zum Höhenjäger Ta 152

Heinz Nowarra – Focke-Wulf Fw 190/Ta 152

Heinz J. Nowarra – Die deutsche Luftrüstung 1933–
1945

Frappe / J.Y. Lorant – Le Fw 190

J. Dressel / M.Griehl -– Fw 190 / Ta 152

Alfred Price – Sie flogen die Fw 190

Gersdorff/Grasmann – Flugmotoren und Strahltrieb-
werke

Kurt Müsegades – Der Militärflugplatz von Delmen-
horst-Adelheide

Wings, Vol. 14 – Restauration of the Fw 190

Rüdiger Kosin – Die Entwicklung der deutschen
Jagdflugzeuge

Wolgang Wagner - Kurt Tank - Konstrukteur und
Testpilot bei Focke-Wulf

Georg Hentschel – Die geheimen Konferenzen des
Generalluftzeugmeisters

Eric Brown – Berühmte Flugzeuge der Luftwaffe

Jeffrey L.Ethell – Monogram Close Up 24 - Ta 152

Christopher Chant – WW II Aircraft

Luftfahrt International 3 – Daimler Benz DB 8-603
B1

Luftfahrt International 5 – Begleitjäger Ta 152 H

Luftfahrt International 10 – Fw 190 Höhenjäger 2

Herzlich Bedanken möchte ich mich für die große
Hilfsbereitschaft und freundliche Unterstützung von
Dietrich Alsdorf, Frank Berger, Robert Braken,
Thomas Bußmann vom Flugplatzmuseum in Cottbus,

Rick Chapman, Herrn Heintzer vom Historischen
Archiv der Mercedes Benz AG in Stuttgart, Richard
Faltermair,

Kurt Müsegades, Stephan Ransom, Christoph Regel,
Gerhard Roletschek, Jürgen Rosenstock, Ursula
Schäfer-Simbolon vom Deutschen Technikmuseum
Berlin, Günter Sengfelder, Konrad Soppa, Rolf Spille,

Christian Stopsack und insbesondere den Ta 152 H
Piloten Armin Mehling und Willi Reschke. Für die
Möglichkeit, dieses Buch mit zahlreichen und hervor-
ragenden Bildern zu unterstützen, bedanke ich mich
bei Herrn Peter Petrick aus Berlin. Herrn Peter Achs
aus Gera danke ich für die vielen Focke-Wulf- und
Jumo-Motoren-Unterlagen, die er mir zur Verfügung
stellte.

Mein besonderer Dank für Ihre Unterstützung gilt den
Ehefrauen der Testpiloten Werner Bartsch, Bernhard
Märschel und Friedrich Schnier.

Für die Gesamtdurchsicht meiner Arbeit und den
hilfreichen Anmerkungen, bedanke ich mich mit
großem Respekt bei Herrn Dipl.Ing. Hans Sander,
Cheftestpilot und Chefingenieur der Abteilung
»Muster-Erprobung« von Focke-Wulf.